An Illustrated Encyclopaedia

WONDERS
of
ANIMAL LIFE

About the Editor

Sir John Alexander Hammerton (1871-1949) was an English statesman and author, best-known for his editorship of popular encyclopaedias published in the first half of the 20th century, such as the 10-volume New Book of Knowledge.

From 1914 to 1919, Hammerton was joint editor with HW Wilson of the periodical "The Great War: The Standard History of the All-Europe Conflict", published by the Amalgamated Press. It was lavishly illustrated with photographs and with realistic drawings which were sometimes hardly distinguishable from photographs.

He was also a contemporary and friend of Arthur Mee, about whom he wrote a biography.

An Illustrated Encyclopaedia

WONDERS *of* ANIMAL LIFE

by Famous Writers on Natural History

Volume Four

Edited by

J.A. Hammerton

Over Three Thousand Illustrations from Photographs and Thirty-three Colour Plates

LOGOS PRESS, NEW DELHI

ISBN -13: 978-81-7268-175-3 (Set)
ISBN -10: 81-7268-175-5 (Set)

ISBN -13: 978-81-7268-176-0 (Vol. I)
ISBN -13: 978-81-7268-177-7 (Vol. II)
ISBN -13: 978-81-7268-178-4 (Vol. III)
ISBN -13: 978-81-7268-179-1 (Vol. IV)
ISBN -13: 978-81-7268-180-7 (Vol. V)
ISBN -13: 978-81-7268-181-4 (Vol. VI)
ISBN -13: 978-81-7268-182-1 (Vol. VII)
ISBN -13: 978-81-7268-183-8 (Vol. VIII)

ISBN -10: 81-7268-176-3 (Vol. I)
ISBN -10: 81-7268-177-1 (Vol. II)
ISBN -10: 81-7268-178-X (Vol. III)
ISBN -10: 81-7268-179-8 (Vol. IV)
ISBN -10: 81-7268-180-1 (Vol. V)
ISBN -10: 81-7268-181-X (Vol. VI)
ISBN -10: 81-7268-182-8 (Vol. VII)
ISBN -10: 81-7268-183-6 (Vol. VIII)

Edition : 2008

Printed & Published by

LOGOS PRESS

Building No. 4788-90, Street No. 23, Ansari Road, Darya Ganj, New Delhi-110 002 (India)

Phone : 23272187

Foreword

Here is not a 'natural history' in the accepted meaning of that description. There exist many and excellent natural histories and more than fifteen years ago the present editor was associated with the publication of one which, in its authoritative letter-press and profuse illustration, had wide acceptance. But it is undeniable that the general reader finds large and exhaustive treatises on animal life, encumbered with Latin names, somewhat hard reading, however fine their illustrations, if they are arranged on the approved scientific divisions—order, family, genus, species—into which naturalists split up animated nature. A proper natural history, however, must follow this method of treatment if it is to form the basis of a scientific knowledge of the subject.

Does the ordinary reader desire such knowledge? Does he require it?

The answer to both questions is a negative: emphatic in the first and modified in the second case. No matter how slightly youth is drawn to the conventional study of zoology, not even the forbidding scientific approach to natural history can rob it of its inherent fascination.

A boundless curiosity about the infinite diversity of creatures that share the life of the globe with him is a deep and abiding and a most natural trait in man. The stories of a Adam naming the beasts, and of the animals, two by two, going into the ark are ancient evidence of man's interest in his fellow-creatures of the so-called lower orders. A study of the life-histories of many animals and birds, fishes and reptiles, or of the astounding social systems of certain insects, is calculated to limit man's pride in himself as the incarnation of the highest intelligence or the finest physical product of the animal world. Especially is it likely to extend his sympathy to all lowly things that are living out their little days with him upon this planet. It lies at the root of humanitarianism. Interest in animal life is never an idle curiosity, and it may prove more absorbing than any other interest that could be named.

Not, then, to add a new natural history to a list already adequate, but to meet this universal and most healthy interest in the astonishing realities and the mysterious ways of the animal kingdom is the purpose of WONDERS OF ANIMAL LIFE. Its title adequately suggests its scope: to enquire into and to illustrate all that is of uncommon interest in the life of the animal kingdom.

While the reader will find in our pages a constantly changing panorama of animal life, with no more apparent order than the naturalist finds in the Brazilian jungle, he can hardly fail to be interested and he is certain to be instructed, for though the only method that is observed throughout our chapters is a rough alternation of interest, now in mammals, next in reptiles, then in birds, fishes, insects, each chapter of this work is written either by one of the leading specialists on his subject or—where expert knowledge is less needed than vividness of descriptive style—by a popular author.

A glance at the list of contributors, their scientific and literary credentials, and the subjects allotted to them in our scheme, will indicate to the reader that though WONDERS OF ANIMAL LIFE is designedly popular in its appeal, it is not the less authoritative in being more entertaining than the conventional sort of natural history.

Never before has there been brought together so representative and so varied a collection of photographic documents covering the whole range of animated nature. In the last decade, and particularly within the latter part of it, naturalists have acquired an immense amount of new knowledge, and marked progress has been made in recording the wild life of the world by means of direct photography.

The exploits of the camera experts employed by the organizers of some of the great 'nature and adventure' films—'Simba' is a fine example—that have fascinated millions throughout the world in these years have produced so rich a body of material for illustrating such a work as this that editorially we are in a more favoured position than any of our predecessors in the same field and we hope the reader will agree with us in thinking that we have taken advantage of that fortunate circumstance for his entertainment and instruction.

J.A. HAMMERTON

Publisher's Note

We are happy to present the illustrated encyclopaedia—Wonders of Animal Life—to the learned readers. As its title suggests, it attempts to enquire into the astonishing realities and the mysterious ways of life of the animal kingdom—mammals, reptiles, birds, fishes and insects.

Presented in eight volumes, each chapter of this work is written either by one of the leading specialists on the subject or—where expert knowledge is less needed than vividness of descriptive style—by a popular author. The text is enriched with a representative and varied collection of photographic documents covering the whole range of animated nature. Our other notable illustrated encyclopaedias include *The Living Animals of the World* (4 volumes), *The Wonders of the World* (4 volumes), *Peoples of All Nations* (14 volumes), and *The Earth and Its Inhabitants* (8 volumes).

TABLE OF CONTENTS

List of Colour Plates

Volume Two

List of Colour Plates

Volume Three

List of Colour Plates

Volume Four

List of Colour Plates

Volume Five

List of Colour Plates

Volume Six

List of Colour Plates

Volume Seven

List of Colour Plates

Volume Eight

List of Colour Plates

Chapter LXII

Animal Slayers of the Snakes

By W. P. Pycraft

Author of "The Courtship of Animals"

SNAKES, the world over, have an evil name: to kill one, anywhere, is deemed a meritorious act. So intense, indeed, is this blind hatred that creatures which, by force of circumstances over which they have no control, have come to assume the semblance of snakes are made to share the same fate, death, whenever the opportunity offers. A girl once told me that though she knew the blind-worm was perfectly harmless, and no snake, yet she always killed one when she saw it, because, forsooth, it looked like a snake. Could any more fatuous reason be offered in excuse for such wanton folly?

It is assumed that all snakes are, and must be, venomous; and that, furthermore, all are viciously aggressive. This, however, is very far from being true, for even those with the most deadly bite will reserve their venom for its legitimate purpose as long as is consistent with their safety.

But the human race, which in so far as the civilized section is concerned, prides itself on its intelligence, is not the only, nor the principal, enemy of snakes, for animals of all sorts kill and eat them whenever possible. Some, indeed, contrive to subsist mainly on snake-meat, and among these the most deadly of all their foes are—snakes!

It seems strange indeed that snake should eat snake, but such is the case. It is not that any snake will fall upon his neighbour and devour him, if he can, but that there are some species which have persistently and of set purpose turned "snake-eaters." The largest and most formidable of these is the hamadryad, or snake-eating cobra, or king-cobra, a species which ranges from India to South China, and attains to a length of twelve feet or more. Its food seems to consist entirely of other snakes. The dreaded krait, which causes more deaths in India from snake-bite than any other snake, feeds chiefly on rats and snakes. It is very common in Bengal and Southern India, and displays an uncanny fondness for creeping into houses, hence the great number of its human victims. The coral-snake of tropical South America feeds chiefly on worms and small burrowing snakes. This species, nearly a yard long, is remarkable for the brilliancy of its coloration, consisting of alternate rings of coral-red and black—a warning-coloration, and a fine example of mimicry, for it is quite harmless as is shown by the fact that, for "cooling purposes," it is sometimes worn by ladies round their necks!

IT may seem strange that snake should eat snake, passing by other possible victims in the hunt for snake-meat. But it is no more so than that bird should eat bird, as do the hawks. Whether the hamadryad eats those of its own species is not known. Probably it does not, unless the adults snap up youngsters that come in their way; for contrary to the general belief, snakes are not immune from snake-venom, even of their own, if introduced artificially into their blood. And this being so, a full-grown hamadryad would probably avoid any quarrel with one of his own kind, though doubtless he would not hesitate to tackle any other poisonous species, since cobra venom is somewhere about sixteen times more powerful than any other.

It is not a little surprising to find that there are many beasts and birds which kill and eat snakes, and many of them are said to be immune from snake poison. The mongoose, the hedgehog and the pig are numbered among these fortunate ones.

Much has been written about the mongoose in this connexion, but a great deal of it is pure guess-work, or based on "travellers' tales." Thus Pliny told his contemporaries that the mongoose, or ichneumon

MARSH MONGOOSE, SLAYER OF ADDERS

Of the various kinds of mongoose inhabiting Africa the marsh mongoose is partly aquatic in its habit of life, and is found, as its name suggests, in swampy places. Besides snakes, it eats frogs and fresh-water crabs. It is famed as a great killer of puff adders.

MOLE SNAPS THE BACKBONE OF AN ADDER WITH ITS STRONG TEETH

While thinking of animals that kill snakes it would hardly occur to many people that so apparently innocuous a creature as the mole should be numbered among them. Yet this remarkable photograph shows a mole in the act of biting through the spine of a viper and doing it with such speed that the snake has no chance to retaliate. It should be remembered that the spinal column of an adder is a very delicate thing and the mole's muscular development is quite out of proportion to its bulk—the average length being about six inches.

first coats its body with mud, in which it wallows, and with this armour can defy the serpent. Topsell improves upon this. "It burrows," he tells us, "in the sand and when the aspe espyeth her threatening rage, presently turning about her taile provoketh the ichneumon to combate, and with an open mouth and lofty head doth enter the list to her owne perdition. For the ichneumon, being nothing afraid of this great bravado, receiveth the encounter, and taking the head of the aspe in his mouth, biteth that off to prevent the casting out of her poison." A more spirited and more accurate account of one of these encounters has been given by Mr. Rudyard Kipling in his story of Rikki-tiki!

Modern observers, like Sir E. Tennant, commenting on the widespread belief that should a mongoose be bitten it immediately rushes off to eat the leaves of some plant which provides an antidote to the poison, remarks: "If the ichneumon"—a name often used instead of mongoose—"were inspired by that courage which would result from the consciousness of security, it would be so indifferent to the bite of the serpent, that we might conclude that, both in its approaches and its assault, it would be utterly careless as to the precise mode of attack.

"Such, however, is far from being the case, and next to its audacity nothing could be more surprising than the adroitness with which it escapes the spring of the snake under a due sense of danger, and the cunning with which it makes its arrangements to leap upon the back and fasten its teeth in the head of the cobra. It is this display of instinctive ingenuity that Lucan celebrates where he paints the ichneumon diverting the attention of the asp by the motion of his bushy tail, and then seizing it in the midst of its confusion."

THAT the mongoose is immune from the poison and is therefore unconscious of any risk attending his attack seems to be shown by a story told by the late President Roosevelt, which was told to him by his friend Tarlton when he was hunting with him in South Africa. He says "Tarlton told me an interesting anecdote of a white-tailed mongoose and a snake. The mongoose was an inmate of the house where he dwelt with his brother, and was quite tame. One day they brought in a rather small puff-adder, less than two feet long, put it on the floor and showed it to the mongoose. Instantly the latter sprang towards the snake, every hair on its body and tail on end, and halted five feet away, while the snake lay in coils like the thong of a whip, its head turned towards the mongoose. Both were motionless for a moment. Then suddenly the mongoose seemed to lose all its excitement; its hair smoothed down and it trotted quickly up to the snake, seized it by the middle of the back—it always devoured its food with savage ferocity—and settled down comfortably to its meal.

SECRETARY BIRD THAT STAMPS SNAKES TO DEATH AND THEN SWALLOWS THEM

Throughout South and East Africa there is to be found a winged slayer of the snakes—the secretary bird. It gets its name from the long feathers on the head which rather resemble quill pens stuck behind the human ear. This crest the bird erects when in conflict with a snake. Its custom is to stamp the snake's life out of it with the powerful legs, and great agility is shown in avoiding the deadly bites, the wings apparently playing the part of shields. The bird stands more than four feet high and is a strong flier.

OPOSSUM HUNTS AND KILLS A SNAKE IN THE WOODS

Deadly as their poison may be, the snakes of America may have cause to regret meeting an opossum. Above we witness an encounter in the woods between an opossum and a mocassin snake. The 'possum stalked its prey round the end of a fallen tree trunk and then jumped swiftly with fatal results for the snake. Below we see the corpse laid across the bole while the opossum prepares to make a good meal out of its dangerous capture. Snakes are, on the whole, delicate in structure and it does not take a great bite to dislocate their backbones.

W. S. Berridge

EGYPTIAN MONGOOSE AND SHORT-TAILED MONGOOSE FROM THE MALAY STATES

Many legends have grown up about the mongoose, and it is supposed to be immune from the effects of snake-bite. But it seems certain that, on the rare occasions when a mongoose, through age or infirmity, does get bitten it dies. The thick fur helps to save the little animal for when this is fluffed out the snake may bury its fangs in hair without even touching flesh. But the mongoose relies mostly on its speed. Besides snakes, the mongoose eats rats, mice, birds and also fruit. Below is the Egyptian mongoose and above the short-tailed species.

Like lightning the snake's head whipped round; it drove its fangs deep into the snout or lip of the mongoose, hung on for a moment and then repeated the blow. The mongoose paid not the least attention, but went on munching the snake's body, severed its backbone at once, and then ate it all up, head, fangs, poison, and everything, and it never showed a sign of having received any damage in this encounter. I had always understood that the mongoose owed its safety to its agility in avoiding the snake's stroke, and I can offer no explanation of this particular incident."

The little hedgehog is a snake-eater. Its method of attack has been watched in captive specimens. When a snake, poisonous or otherwise, is placed in its enclosure it is found that the hedgehog's plan is to snap quickly, and then, instantaneously to present its spiny back to the reptile, which, if poisonous, strikes at the spines and expends its poison uselessly. The harmless slow-worm is dealt with in exactly the same way, showing that the attacker takes no chances!

The battles between hedgehogs and snakes have been recounted again and again, from ancient times. Topsell gives a picturesque account of such an encounter. "There is," he says, "mortal hatred between the serpent and the hedgehog, the serpent seeketh out the hedgehog's den, and falleth upon her to kill her, the hedgehog draweth itself up together, round like a football, so that nothing appeareth on her but her thorny prickles; whereat the serpent biteth in vain, for the more she laboureth to annoy the hedgehog, the more she is wounded and harmeth herself, yet notwithstanding the spite of her minde, and hate of her heart, doth not suffer to let go her hold, till one or both parties be destroyed. The hedgehog rowleth upon the serpent, piercing his skin and flesh (yea many times tearing the flesh from the bones), whereby he scapeth alive and killeth his adversary, carrying the flesh upon his spears, like an honourable banner won from his adversary in the field."

The armadillo is a snake-eater, and is said to kill his victim by contriving to get the body under his shell, and then moving backwards and forwards so that it is sawn in pieces by the sharp edges of this bony shield. In the process the victim makes frantic efforts to wound his antagonist, but the bony armour is proof against the serpent's tooth. As soon as it is slain the victor proceeds to eat the vanquished, commencing at the tail.

The glass snake is another snake-eater. This creature, by the way, is not a snake, but one of the limbless lizards found in the Balkan Peninsula, South Russia, Asia Minor and Morocco, and runs to more than a yard in length. Other species live in the Himalayas, and Burma, and North America.

Its food is, indeed, varied, though its staple diet seems to consist of snails, the shells of which are crushed, and of mice; but nothing that can be seized comes amiss, worms, insects, young birds, small lizards and vipers are all eaten with a relish. The larger victims it encircles with a lightning twist of the body, after the manner of the constricting snakes, or they are shaken, as a terrier shakes a rat, till dazed, or stunned, when they are broken up by the powerful jaws and swallowed in pieces. The heavy armature of the body, formed by the scales, makes the glass snake invulnerable to the bites of the viper, should the latter have any opportunity of making a fight for life.

Though its bite is so formidable, it is to be noted that the glass snake when caught makes no attempt to use its jaws, but instead, twists itself round one's hand and arm, besmearing them with a disgustingly stinking fluid which it discharges. Nevertheless, it is readily tamed.

It seems but just, since there are some snakes which eat lizards, that there should be some lizards which eat snakes. The eyed-lizard, common in the South of France and the Riviera, is one of these. It is a voracious animal and will attack any small lizard, slow-worm, or snake it can find. This species, by the way, attains to a length of as much as two feet and is readily tamed in captivity. The teju (*Tupinambis punctatus*) of South America, which attains a length of three feet, is another snake-eater. It is much disliked near homesteads, since it is given to robbing hen-roosts, taking both chickens and eggs. As a consequence it is hunted down by dogs, and when killed is generally eaten by the hunters, its flesh being much esteemed.

Many birds eat snakes whenever opportunity offers. And of these the most famous is the South African secretary bird, though some discount its prowess in this regard. Let us take the evidence first of Jules Verreaux. "The secretary bird," he remarks, "has been modelled on a plan appropriate to its mode of life: and it is therefore for this purpose that, owing to the length of its legs and tarsi, its piercing eye is able to discover at a long distance the prey which, in anticipation of its appearance, is stretched on the sand or amongst the grass. The elegant and majestic form of this bird becomes now even more graceful; it now brings into action all its cunning in order to surprise the snake which it is going to attack; therefore it approaches with the greatest caution. The elevation of the feathers of the neck and back of the head shows when the moment for attack has arrived. It throws itself with such force on the reptile that often the latter does not survive the first blow. But if the bird does not succeed and the enraged snake draws itself up, and expands at the same time the skin of its neck, as is the way with the more dangerous serpents, the bird is forced to retreat, and takes a spring backwards, waiting to seize a favourable moment for recommencing the attack. Raising itself, the furious reptile moves its tongue with the quickness of lightning, and gives forth the most vehement hisses, which keep back the enemy and seem to force some

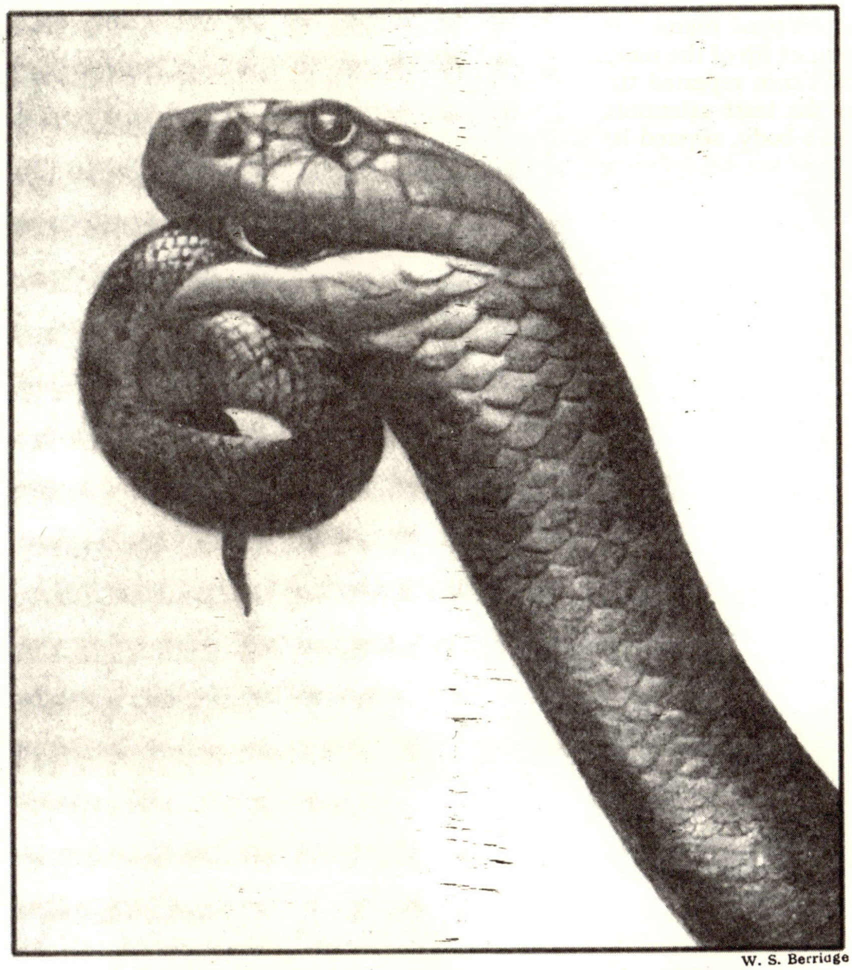

W. S. Berridge

WHEN SNAKE SLAYS SNAKE: THE KING COBRA

One of the most determined destroyers of snakes is the king cobra or hamadryad of India and Southern China. It grows to a length of twelve feet, and moves with great speed. We see one here swallowing another snake, which has coiled itself in the attempt to make a loop too large to be engulfed. But the hamadryad's lower jaw is slowly forcing the coil apart.

and as soon as it arrives at the head, it completely smashes the skull by several blows of its bill before swallowing it."

Some eagles, as well as harriers, eat snakes. The serpent-eagles (*Spilornis*) of the Indian and Malayan regions are snake-eaters, as also are their near relations the buzzard-eagles (*Butastur*). Still more addicted to snake-eating are the South and Central African bateleur eagles, which attack both venomous and non-venomous species, be they large or small; the birds rapidly disabling their prey with their formidable beaks. When the grass-jungles are on fire these birds beat along the line of flame and seize the snakes and other reptiles as they creep out, sometimes dashing into the smoke for this purpose. The American swallow-tailed kite is a confirmed snake-eater, and has been seen sailing leisurely along and from time to time bending its head down to tear a piece from a small snake it held in its talons.

Jungle fowl and peacocks will eat snakes, though there are no records to show whether these are ever of poisonous species; and the same is true of some of the storks, including the adjutants.

respect from it: but the bird, whose courage redoubles in the same ratio as the difficulties increase, opens out its wings, and returning to the charge, assails the reptile afresh with blows from its terrible feet, such as no one would believe, and which are not long in putting the snake *hors de combat*.

"We have, however," he continues, "sometimes seen the snakes launch themselves on the secretary, but, either by opening its wings, whose long primaries serve it as a kind of shield, or by jumping backwards, or on one side, the bird is certain to parry the attack of its antagonist, who, at last, overcome by fatigue, falls at full length on the ground. The moment is seized by the secretary bird to redouble its massive blows which, by dislocating the vertebral column, soon cause the animal to give up the ghost. It is then that the victorious hawk darts like an arrow, and placing its foot upon the serpent's neck, just at the back of the head, commences to swallow it, which it does by beginning at the tail. Nor is this a long operation, even with reptiles five or six feet in length and more than four inches in diameter;

The importance of the alligator as an agent in keeping down the numbers of the deadly copper-head or mocassin snake in Florida was not realized until the numbers of the alligators became seriously diminished by the high prices offered to hunters for their hides. As their numbers decreased so the numbers of the snakes went up, furnishing us with yet another illustration of the evil effects of ignorance in disturbing the balance of Nature. It may, perhaps, be thought difficult to regard the alligator with any great increase of partiality on the grounds that it destroys poisonous snakes. Indeed, one might feel that the snakes were the lesser of two evils. Nevertheless this balance of nature is very delicately adjusted and the results of upsetting it have often gone beyond—alarmingly beyond—expectation.

Lastly, in connexion with the snakes which feed on snakes it is interesting to turn to page 140 where the bottom photograph shows us a non-venomous grass snake devouring a viper. Yet there are people who kill grass snakes on sight just because they are snakes and some snakes are poisonous.

FIFTY THOUSAND RACING PIGEONS START UPON A GREAT FLIGHT FOR HOME

Pigeon racing is founded on the amazingly powerful instinct of the birds to fly back to their homes. The sport was introduced into England in the early 'seventies and, especially in the North, has accumulated many enthusiasts. The method is to despatch the competing birds, which are specialised as either long or short distance performers, to an agreed starting place at a distance from the home of each bird commensurate with its powers. Then the competitors are released and the time is registered. At the other end of the journey each owner is waiting, and so soon as his bird arrives a rubber ring is removed from its leg and is placed in a special registering clock. The clocks are then inspected, and the bird showing the highest speed is judged the winner.

Chapter LXIII

The Homing Instinct

By H. Mortimer Batten

Author of "Habits and Characters of British Wild Animals"

THE homing instinct in Nature is one of those problems which will ever provoke a sense of wonder in mankind. Man, with his knowledge of mathematics, his instruments, his knowledge of the stars, can grope his way across seas and continents, but in the migration of birds, beasts, insects and fishes, the travel sense amounts to an inbred knowledge of direction backed by an infallible and wonderful sense of the seasons.

Without touching upon migration, however, let us take the homing instinct in its simplest forms. Here is an example. Mr. Fisher, who lived at Blairghour Farm on the shores of Loch Awe, possessed a sheep dog. This was before the days of railways in that vicinity, and during the Glasgow sheep fair Mr. Fisher took his dog with him to that city. They went by road and by boat, and the dog, a young animal, took the fancy of a Glasgow butcher, who offered so high a price for it that Mr. Fisher accepted, though he had no special desire to sell the dog. He then returned home to Loch Aweside, but to his utter surprise the dog, all by itself, arrived back from Glasgow a few days later.

Now, a glance at the map will show what this meant. For the dog to return by the way it was taken would have been impossible ; it must have returned by land, which meant an enormous circuit of the west coast lochs, through country over which it had never previously travelled. It must have covered a distance of well over sixty miles, going far north of its destination, then turning south to gain its home. Thus the dog's homing sense did not merely lie in covering a straight line from point to point ; it meant far more—the direction of home after it had exceeded the distance in a northerly direction, and was compelled to travel back.

OF all wild creatures of Britain, the homing instinct is probably more strongly developed in the red deer than in any other. The red deer have a great love for their own native ranges—for the land in which they were born, and where their mothers nursed them. A stag in his later life may cross ranges and rivers, but always, when no other instinct calls, he returns to his native land. Much the same applies to hill sheep. I have known a horned ewe to cross two ranges in order to regain her native hills from which she had been sold.

Sometimes, to improve the blood, deer are exported to distant lands, or maybe from the mainland to some outlying island. The stag is pounded and probably netted out, and amidst the general turmoil, his horns are sawn off. He is shut in a crate, from which he can see nothing, and the crate is eventually loaded aboard a motor van, which, by intricate hill roads, finds its way to the loch steamer. It is taken aboard the steamer, probably by means of a hoist, steadily revolving till it reaches the deck, and the boat gropes her way out by the tortuous routes of the sea. Thus the stag is taken to some new locality and liberated, but a week or so later he is observed back on his native range. Scottish stags exported to New Zealand have been found far out at sea, swimming faithfully towards the North—heading, oblivious of certain death, across the width of the world for their native Scottish hills !

WHEN I was a youngster I was well acquainted with a pet roe-deer. When he was small he was a delightful little companion, and dearly loved his little mistress. No child could have had a daintier or more attractive pet, but as he grew older he resented any trespass upon his little mistress's companionship, and his budding horns were becoming dangerous weapons. During his life he had never so far ventured beyond the gardens and policies of his home, where his needs and amusements were met, but on his becoming mischievous, it was decided that he must be returned to his own wild race among the mountains. He was, therefore, trussed and boxed up in a laundry hamper, which was carried across the ranges in a ramshackle old motor-car, then by boat across a loch to a keeper's lodge on the other side of the range. Here he was handed over to the keeper who was in charge of the outlying beat, it being thought that the man might wean him gently from the ways of domesticity until he was ready and glad to gravitate back to his own kind once more.

At the keeper's lodge the laundry basket was opened, at which the little buck, red-eyed and indignant, bounded out. In the twinkling of an eye he had scattered the spectators, then over the wall he went, and into the birch woods. That evening, within an hour of the motor-car arriving back, the roebuck's little mistress heard the familiar tapping of his hoofs on the door of the summer house, where she was accustomed to do her lessons.

A friend in Edinburgh tells me that in the days of his boyhood the family had a much beloved retriever attached to their summer residence at Gartincaber, and during the summer they became so fond of him that it was decided to take him home to Edinburgh for the winter months. The journey was made by road, but the old dog was shut up in the rumble of the equipage, from which he could see nothing, and was probably not too liberally supplied with air. Thus he was conveyed across Scotland, but no sooner was he liberated in Edinburgh than he fled in panic. Late the following day he was back at his native kennels at Gartincaber.

Of course, a multitude of questions arise. Probably the first conclusion to which we come is that in these

HOMING PIGEONS IN FLIGHT FROM HITCHIN TO THE NORTH

Here is part of a great flight of over forty-six thousand pigeons which are engaged in a race from Hitchin, Hertfordshire, to their homes in the North of England. The birds, packed in some two thousand baskets, arrived in two special trains from South Shields and Newcastle, picking up competitors as they went. The lower photograph shows a racing pigeon.

cases a dog backtracks by scent. This may be possible on occasions, but not always. There are many cases which disprove it, such as the case of the dog which found its way back to Loch Awe. Here is another. A gentleman I know lived about eighteen miles from Paris. He went to Paris by train each day, crossing the city from east to west. Sometimes he took his little spaniel with him, and one day, when he had left the dog at home, he was surprised in the late forenoon to find that young gentleman ushered in by a clerk. The little dog greeted his master with delight, but the mystery remained. How did the animal find his way across country and across the city of Paris to the offices?

The matter did not end there, however, for a few days later the door of the office was again opened, and this time not only the spaniel frisked in, but also a bulldog and a Dalmatian hound, dogs which belonged

J. J. Ward

RAT RETURNING TO ITS HOLE AFTER A PROWL ROUND THE GARDEN

However far it may roam in its nightly, and sometimes, daily wanderings in search of food, the rat has one home and returns to it until circumstances force it unwillingly to seek a new abode. In the upper photograph a common brown rat is prowling about the garden and is examining the rock border which separates a path from the flower bed. Below we see it beside the hole it has dug for its home, unheedful of the unquestionable publicity of such a site, right in the middle of the garden path.

J. J. Ward

POSSESSING A STRONGLY DEVELOPED HOMING INSTINCT: THE HEDGEHOG

In the thickest part of a hedge or in some dense undergrowth the hedgehog makes its home and spends the day-time there rolled up and asleep. When night comes the little mammal wakes up and goes out to seek its food, which may be anything from an adder to a plant root. If kept in a semi-tame condition in a garden the hedgehog is a useful destroyer of slugs and, indeed, except to those who preserve game, the animal is entirely harmless despite old countrymen's prejudices to the contrary. At dawn the hedgehog goes home.

to the stables of my friend's establishment, and which had never previously visited the city.

In the far north-west some marvellous stories are told of the homing instinct in dogs of the husky and malemute breeds, but many of these accounts are difficult to prove. I remember one touching case, however. A prospector died away up-country, about three hundred miles from Nome City, and was lowered into the ice by his companions. His favourite malemute would not reconcile himself to any other master, but when the ice broke four months later he boarded the river steamer with a throng of adventurers and went to Nome, which the dog had often visited with his master. There for several weeks this great husky brute searched the drinking dens and the billiard parlours, meeting every canoe which came in from the creeks till he became quite a well-known figure. What became of the dog eventually I do not know.

THE homing instinct in dogs is little more than we might expect in view of their devotion and intelligence, and parallel examples would be hard to find among the rank and file of our other fireside pet, the cat. Still, cats are not lacking in this wonderful respect, as is proved by the following example coming within my own personal experience.

One day a picnic was arranged for a party of children, and one child wished to take her kitten with her. The kitten was packed into the usual double-topped hamper, and the party went by car, crossing the river by the one and only bridge for several miles, then through the birch woods on the south side of the loch. The picnic was held about seven miles from home, so that between that place and the point at which the kitten was liberated there extended a rapidly flowing mountain river, which emptied itself into a loch several miles in length. The only way back, then, was across the bridge, and having regard to the roughness and the wildness of the country, one would hardly expect a small animal like a kitten to be able to find it.

Well, the kitten was liberated from the basket, and enjoyed the picnic as much as anyone, but when it came to returning home, no kitten could be found. The search extended till dusk, and in the end a very forlorn party was compelled to abandon the hunt. The following morning the kitten arrived home—wet and bedraggled and hungry and wretched, but nevertheless glad to get back.

H. Mortimer Batten

DEER THAT WILL FACE GREAT ODDS TO REGAIN ITS NATIVE HEATHS

Of all wild animals the deer seems to have the homing instinct most strongly developed. Often in Scotland red deer are removed from one district to another for the purpose of improving some local stock. A stag may, when captured, be driven many miles in a motor lorry, taken across a loch in a steamer, and perhaps eventually landed on an island. Yet, in a week or so, the animal will be back again on its old range. Scottish deer taken to New Zealand have been found at sea, swimming steadily in the direction of home.

While certain breeds of pigeons are famous for their homing instinct, their "bump of locality" being abnormally developed, and while this faculty has been increased by breeding for long-distance flight, there can be no doubt that many wild birds are more remarkable than the pigeon in this respect.

It would seem that in the brains of most animals there exists a little compass point bearing infallibly upon the region they love. Distance cannot rob it of its magnetic pull; they may have to turn and yet turn back, but the compass point does not fail. A single error might throw them miles out of their reckoning, but the error simply does not occur. A sense of direction, and that only, is the guiding power. It cannot be scent or memory of landmarks—that is manifestly disproven, and man can, to some extent, acquire the same instinct.

At one time of my life, when for weeks and months on end we travelled by maps and by compass, I found myself acquiring a sense of direction unknown in the ordinary course of life. I acquired the habit at night-time of pointing out the north, and often it was difficult to decide which was the north until one closed one's eyes; then opening the flap of the sleeping-bag, one could check the direction by the familiar glare in the heavens. Thus acquired, this sense of direction has never left me, and in the Underground Railways I can pick out the north.

Take a Red Indian, blindfold him, turn him round half a dozen times, then give him a few seconds to collect himself, and he will lay his fingers on the points of the compass. I have known Indians to travel incredible distances over country they had never previously traversed, sometimes through dense undergrowth where no landmark was visible, yet eventually striking a certain point without a single readjustment of their bearings.

What a savage can learn a white man can acquire if it is worth his while, and there is no doubt that the homing instinct in animals has, up to a point, its counterpart in human beings. In our artificial lives we have lost many gifts and senses, since they are no longer of use to us, and thus the homing instinct in animals is a far greater wonder to us than it is to less civilised people. In civilized man the homing instinct, if he has one, seems to have become rather a sentimental attraction for the place where he grew up and spent his youth, than a definite idea of how to get there in terms of direction.

E. Step

EXTRAORDINARY VARIETY BOTH IN SIZE AND MARKINGS AMONG THE BRITISH SNAILS

In the **left-hand** photograph we have a number of examples of one of the most commonly seen British snails, Helix nemoralis. The shell is ornamented with a series of chocolate-coloured bands, and the variety of the arrangement of these bands is extraordinary among individuals. The ground colour may be almost white, dark brown, flesh-pink, or even bright yellow. This snail should be looked for in hedges and can otherwise be found in very many gardens. It is, with the exception of its near relative, Helix hortensis, the most brightly tinted and variously marked of all British snails. Birds of all kinds, from chickens to thrushes, are very fond of it. The right-hand photograph shows some remarkable variations in size among some of the other land snails.

Chapter LXIV

The Strange Story of Slugs and Snails

By Wilfred Mark Webb

Author of "The Eton Nature Study Note Book"

IN the garden are things with legs—spiders with eight, centipedes with many—insects with wings, and worms which wriggle, but in the current language of the advertisement writer, the snail is the creature "which is different."

In the first place, like the slug, whose name proclaims its habit, the snail moves sedately with the help of the whole underside of its body, which adheres to the ground. Its eyes have not only started out of its head, but are on the tips of sensitive tentacles which can be pushed out some distance in advance of it. The snail differs, however, from its relative in that its large coiled shell prevents us from thinking of the division of its body, by an imaginary line, into two similar halves.

To see how such animals really do get from place to place a piece of glass is necessary, and when a snail, for instance, has started to crawl on one side of this, it can be turned over. Then a remarkable series of ripples, due to the expansions and contractions in the muscular underside of the creature, which is called "the foot," are very evident and result in the front end being slowly pushed forward and the hind pulled after it.

Shells have had an attraction for human beings from the times of the cave men onwards, and are worthy of some detailed study. The spiral ones can be mathematically described as a coiled hollow cone, and, like screws, may be right or left-handed. Those of the common snail and the majority of univalves, both in the sea and on the land, belong to the first category, and when the tip of the shell is pointing away from the onlooker its mouth is on the right side. Left-handed shells, with the spiral reversed, occur as rarities in right-handed species, but there is a number of snails in which they are the rule instead of the exception and a few in which one form of twist is as common as the other.

THE making of a shell is much like the building of a concrete wall. There are three layers, the outermost thin, consisting mostly of animal matter and secreted by the edge of the bag or "mantle" which lines the shell. This horny layer gives the fresh appearance to the shell and during its construction corresponds to one set of boards used for wall-making, while the surface of the mantle makes the other, and the edge again pours out a mass of limy matter into the space, just as the builder would put in his mixture of stones and cement. This calcareous secretion crystallises and forms the second layer, to which a third, similar to the mother-of-pearl in an oyster, is added by the surface of the mantle.

In the sea and in fresh water many of the snail's relatives, like the periwinkle, can shut up their houses, when they retire into them, with a shelly lid permanently attached to the end of the body. A few amphibious forms, and others that have come to live on land more recently than our more familiar species, still possess one. Some among the true snails make a temporary lid in summer to keep themselves from being dried up and in the winter as a protection against the cold or living enemies which cannot manage to break their shells.

Quite substantial structures may be produced as in the case of the so-called Roman snail which uses much lime. This hardens into a smooth white temporary lid, which is thrown away when the spring comes. All sorts of devices have been developed to keep out other carnivorous snails or the grubs of beetles, such as those of the glow-worm, and of other insects.

THE mouths of some shells are made more narrow if the creature lives to become adult, and this in desert snails would also tend to prevent undue evaporation. Pleats and prominences called "teeth" serve commonly to protect the entrance and at times a deep dent on the outside of the shell a little distance behind the opening marks the spot where an extra large internal projection partially blocks the passage.

A further development is found in the high-spired left-handed forms which I have called swing-door snails. Here, in addition to folds in the shell, is an additional safeguard in the form of a curved plate of the same material as the shell hinged to it, which serves to block the entrance when no portion of the animal is outside the shell. This can be seen in position by cutting away a piece of the shell, or a picture of it can be obtained by means of the X-rays.

We may turn again now to the animal. The horns or tentacles have already been mentioned, and of these there are usually four, though in some of the tiniest snails only the upper pair are developed which bear the eyes. These are very easily and quickly put out of harm's way, for their supports are hollow and have a special muscle running up to the top inside. The slightest touch causes the muscle to contract and the eye is pulled into the head in the same way as a glove finger is turned outside in. On the other hand the eye is extended by the contraction of the walls of the tentacles which turn it right side out again. A few species, including the needle snail of Britain, which live underground in the dark, have become blind.

The mode of feeding in snails and ordinary slugs is uncommon to say the least of it. When the mouth is applied, for instance, to a cabbage leaf, as it often is, the animal gets a purchase upon it by applying to the surface a horny-looking "upper jaw" with ribs on it which prevent it from slipping. Then a muscular lever standing up from the floor

F. W. Bond

AMPHIBIOUS APPLE SNAIL AND ITS EGGS

South and Central America are the homes of the amphibious or apple snail, which has a remarkable breathing apparatus enabling it to live as well in water as on land. The long tube used under water is seen extended in the lower photograph. Above are the clusters of eggs this snail lays on water plants. At a distance they rather resemble grubs crawling on the leaves.

of the mouth brings into action a band-shaped flexible rasp disposed over the front and top of the muscle and running up the back from a little bag, in which more of it is being formed and furnished with new " teeth " to take the place of those which are worn out in front. The forward movement of the lever brings the natural rasp through the open mouth into contact with and along the leaf. Portions of the latter are thus effectively removed in a finely divided state ready for digestion, and carried upwards and pushed towards the creature's gullet by the backward stroke of the muscle.

The carnivorous slugs have the same apparatus, but it and they have—as we shall see—been greatly modified in connexion with the capture of their prey and their peculiar underground life.

The breathing of a snail or slug is curiously comparable to the airing of a room by throwing open the window. If one of the animals is watched when crawling, a large opening in the

right side of its body will be seen or will after a time make its appearance. This corresponds to the window. In the snail it is under the edge of the shell and best seen from the underside when the animal is crawling on glass. In most slugs it is near the margin of the shield, in various positions in different genera. In English kinds it is in the front in the great arions of the field paths and their smaller and destructive representatives in the garden whose name begins with A. In geomalacus, the spotted Kerry slug which Ireland shares with Spain and Portugal, it is farther back. When we come to Limax, to which most of our species, the great grey, the yellow, and the field slug belong, the breathing orifice is behind the middle of the shield, and the "T" of testacella, the worm-eating slug, serves to remind us that the opening is not far in front of the tail.

In all cases it leads into the lung chamber (corresponding to the room), whose sides are

F. W. Bond

EDIBLE SNAIL ON ITS WAY TO FEED

While all British snails are probably quite edible—and a snail is presumably no less so than a winkle—only the largest is called the edible snail. Its official title is Helix pomatia, the second word being derived from the Greek and alluding to the lid with which this snail closes its shell. It is said that it was introduced into Britain by Roman soldiers, but this is very doubtful.

Garden snails feeding on rhubarb leaves

Garden snail with shell which it has mended J. J. Ward

Garden snail with its eggs

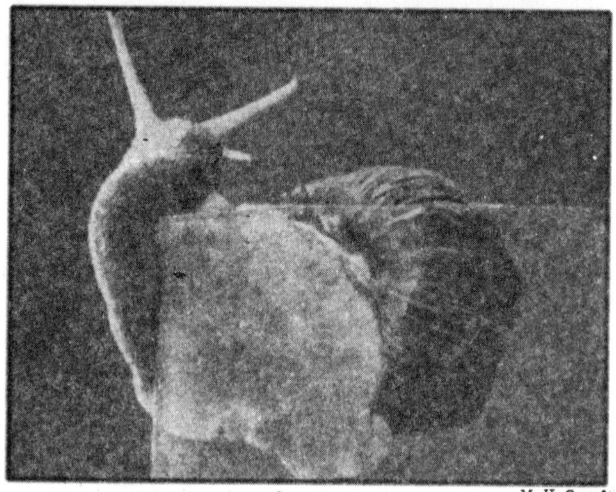

Garden snail travelling over a glass edge M. H. Crawford

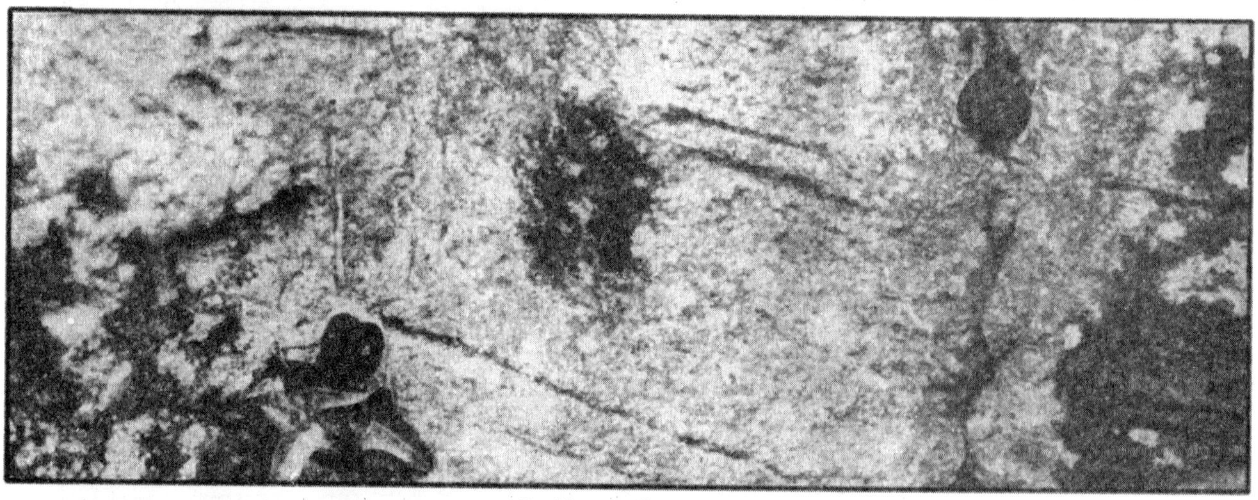

Lapidary snails on a rock E. Step

EVERYDAY ACTIONS IN THE SNAIL'S SLOW-MOVING LIFE

One of the most remarkable things to be noticed about a snail is its method of eating. When the jaws are opened a wonderfully fashioned tongue, rather like a rasp, comes into play. The top left-hand photograph shows Helix hortensis, which is not so often seen in gardens as its name suggests, feeding on a rhubarb leaf. The snail goes to work with a side-to-side movement of the head and the rasp-like tongue cuts the leaf into fine pieces. Another photograph (centre right) shows a snail on a glass plate and the action of the "foot" in moving.

C. W. Bond

J. C. Johnson

J. J. Ward

POND SNAILS AMONG THE WATER-WEEDS THAT ARE THEIR HOMES

Nearly all snails lay eggs, and here is a clutch of water-snail's eggs (bottom left) adhering to the leaves of a water-plant. Like the eggs of the apple snail in page 686, it will be noticed that these look like some grub or caterpillar when seen in a mass. The water-snail in the bottom right-hand photograph looks, from this angle, like some strange new animal. Of the British snails 46 out of the 127 species live in water. The upper illustration shows one of the more common ones, Limnaea stagnalis, which is the largest of its family.

F. W. Bond

the snail a love dart of its own. This is a sharp calcareous weapon, often delicately chiselled or of elegant shape, carefully kept in an inside sheath or pocket, and it is by the turning of this inside out that the point is bared for action. This process might result in the dart becoming detached, and certainly one has been found sticking in a leaf which it had pierced ; but it seems that the suggestion that the dart is actually shot at a venture through the air is a flight of poetic fancy and that the spicule usually remains attached to the owner as no aim could be taken. At the same time it is wonderful enough in itself, albeit it is customarily used more in the fashion of a dagger and in reality acts as a spur. In higher animals in which the sexes are distinct the female only in rare instances does the courting, and therefore it is all the more interesting to note that the love dart of the snail belongs to its feminine side.

Nearly all slugs and snails lay eggs, and those which are most likely to be seen in

plentifully supplied with blood vessels connected with the heart, and through their walls oxygen is taken in and carbonic acid gas given out. When the air in the chamber contains too great a proportion of the latter the orifice opens and by diffusion the normal proportion of the gases in the chamber is regained.

Not the least strange feature of the creatures with which we are dealing is the fact that they, like many plants, have the functions of both male and female sexes carried out by the same individual and at the same time. In these hermaphrodites pairing duly takes place and eggs of each animal may be fertilised by the sperm cells of the other.

Snails are proverbially slow and if a specially placid one, dwelling in the garden or hedgerow, is unresponsive to a would-be lover and returns not the proffered caress, the suitor does not need the kind help of Cupid, for Dame Nature has given to

E. Step

GARDEN SNAIL AND AN APPLE SNAIL'S EGGS

Most snails live from between two to five years. In hot weather they cover up the mouths of their shells with mucus until rain shall fall again. Only a few British snails harm garden plants. Below is a garden snail and above apple snail's eggs magnified

F. W. Bond

England are the semi-transparent ones of the ordinary vegetarian slugs which may often be found lying on the ground under damp matted grass looking like a heap of small pearls without any of their fine lustre.

In the coverings is little or no lime, but in the case of snails this is present, so that they can rightly be called egg-shells. The giant forms of the West Indies and South America produce eggs which might easily be mistaken for birds' eggs. The most familiar kind are as long, if not as broad, as sparrows' eggs, with a pure white matt surface, but there are bigger ones laid by larger species in the last-mentioned place and South Africa which are highly polished and therefore more liable to take people in, and there is one in the shell gallery of the British Museum at South Kensington which is one and three quarter inches long and rivals a pigeon's egg in size.

W. S. Berridge

GIANT SNAIL AND ITS EGG: BRAZILIAN LAND SNAILS

The giant snails of the West Indies and South America lay eggs as long, if not as broad, as those of a sparrow. The shells contain lime and closely resemble birds' eggs in appearance, having a pure white matt surface. Here (bottom) is one of these giant snails with an egg, while the upper photograph shows some specimens of Brazilian land snails.

Some of the swing-door snails hatch their eggs inside their shells, and the fact that the so-called fresh-water whelk does the same thing is indicated by its Latin name of *vivipara*.

The snails belong to various families, and slugs without an external shell are to be found in more than one of these. They gain advantage over the larger snails, for they can creep into narrow crevices out of the way of feathered and other enemies where it would be impossible to drag any but the most minute kind of shell.

The main bodily difference is that most of the viscera, instead of being in a bag which forms a coiled

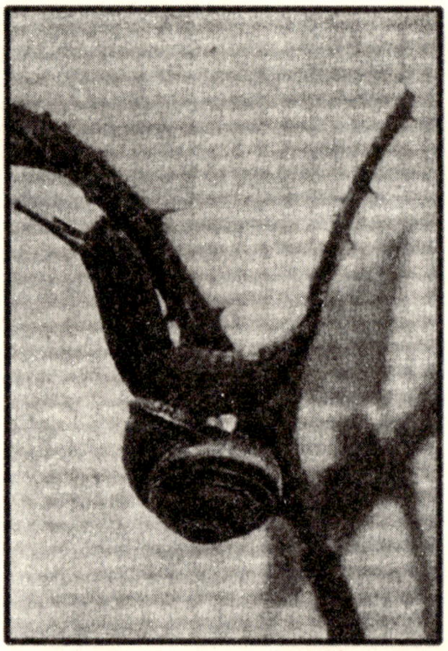

M. H. Crawford

HOW SNAILS OVERCOME THE PLANT'S PROTECTIVE MEASURES

Plants have evolved a number of means by which to protect themselves from the attacks of various enemies, of whom the snails are among the most formidable. The snails, on the other hand, have taken counter-measures, as we see from the above photographs. On the left is a snail climbing the armed stem of a briar by covering the harmful surface with slime. When resting (centre) it hangs by the same slime. It can cope equally well with a thistle leaf, as the right-hand photograph shows.

hump on the back of a snail, is in the hinder end of a slug, and the bag is reduced to a shield, under the front of which the head can be pulled and which in some cases is strengthened by an internal shelly plate. In others, where the skin is extraordinarily tough and unpalatable, as in the black slugs we see openly crawling about after rain, the shell may be reduced to a few limy granules, for these creatures are not often attacked by birds and their vitals need not be specially protected from a peck.

The worm-eating slugs, as we have already pointed out, differ so markedly from the usual run of slugs that they are worthy of special description. They usually have a small ear-shaped outside shell, and this is nearly at the hinder end of the animal. Most of the .nternal organs consequently lie in the part of the body which in the snail we should call the neck.

The worm-eating slugs burrow under ground and crawl along worm holes. The small shell would in such places protect them from an attack from behind. These remarkable creatures are able to elongate their bodies in a way that recalls a cat, and can proceed along a narrow tunnel. On the other hand they may hunch themselves up and become short and thick.

Perhaps the strangest thing about them is the immense development of the muscle in the mouth,

J. C. Johnson

M. H. Crawford

SNAILS SLEEPING THROUGH THE WINTER IN THEIR SHELL HOUSES

Carrying their homes about with them wherever they go, the snails have only to find some damp spot where they can be inconspicuous and there take up winter quarters. Lack of moisture is all they fear, otherwise life is very simplified compared with the lot of many other creatures. Some snails have a lid or operculum with which to close the entrance to their shells, while others effect the same purpose with a layer of slime The right-hand photograph shows a group of eight specimens of Helix pisana.

Red slug crawling over the ground

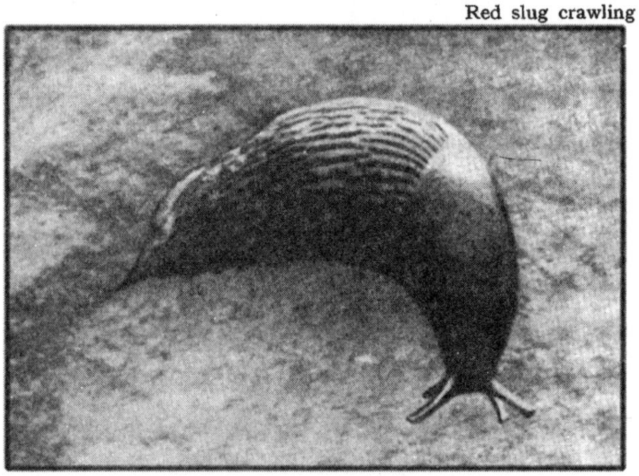

Common black slug

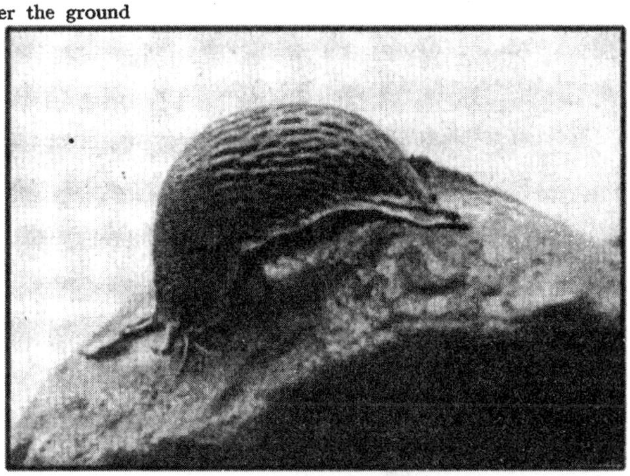

Black slug with body partly extended

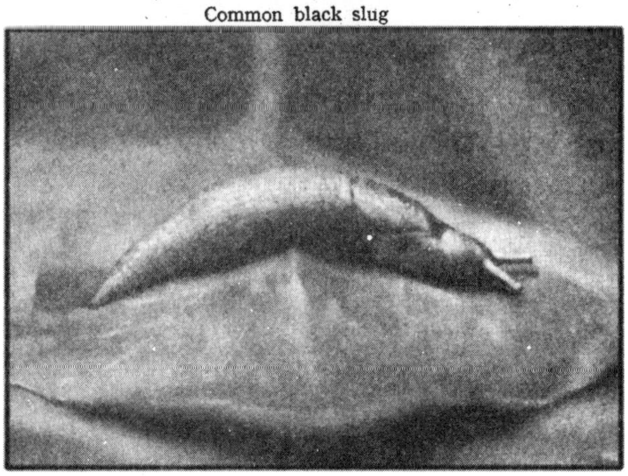

Grey slug

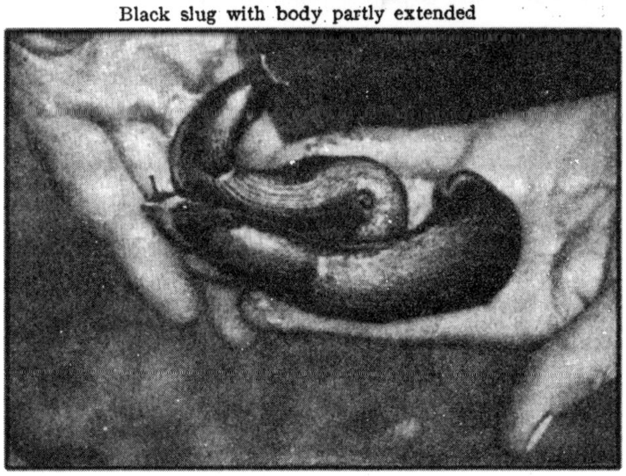

A pair of red slugs

SLUGS, LAND MOLLUSCS WITH HIGHLY MODIFIED SHELLS

In the slugs, what corresponds to the shell of the snail is reduced to a mere shield under which the head can be drawn in case of emergency. In the group, Arionidae, of which we have an example above in the centre photographs of Arion ater, the black slug, the shell is reduced to a few chalky grains beneath the mantle. The black slug relies on the fact that its skin is tough and unsavoury. The "black slug" may be yellow, brown, or red. The grey field slug (bottom left) is common all over Britain. Photographs by Bond, Ward, and Berridge

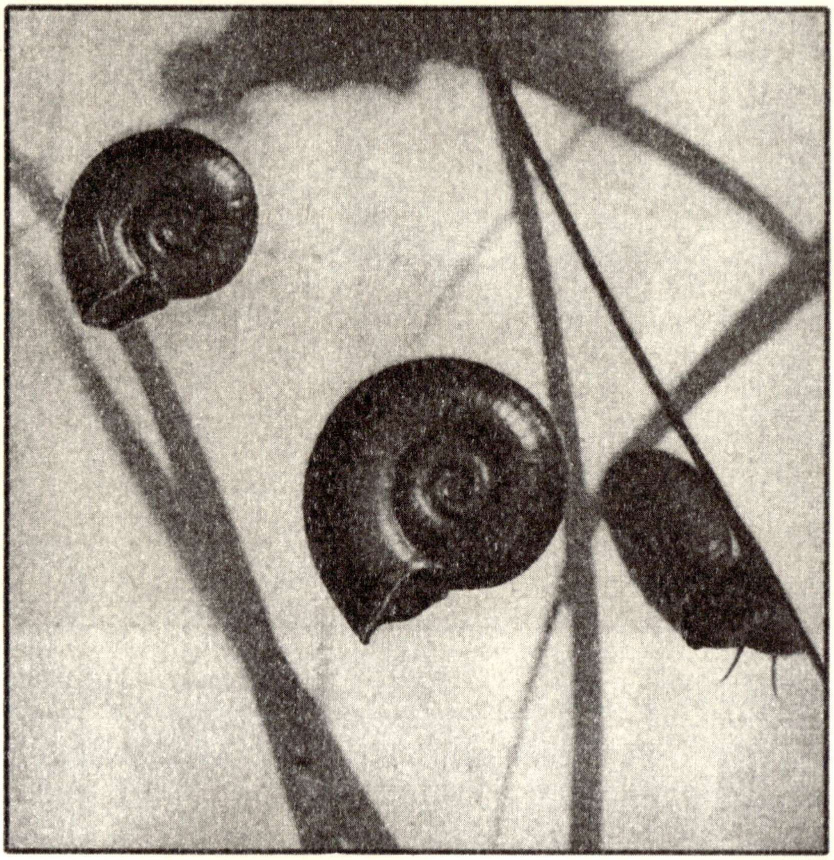

F. W. Bond

When once the worm has been secured, the rest is easy, for as the slug extends itself once more, the victim is gradually drawn into its body and eventually disappears.

The eggs have very thick shells, and are pointed at each end like those of a dabchick. They are large compared with the size of the animal, and although they are not bigger than a barley grain, a careful manipulator has been known to blow them with his mouth.

The worm-eating slugs are by no means the only carnivorous forms, for many other snails belonging to the same family—but with shells of many shapes and sizes which cover them—have also abandoned a vegetarian diet. A few others may also be mentioned, including the largest of the British glass-snails, which is also noteworthy on account of its blue skin. These forms can also elongate their bodies so as to reach up inside the shells of their victims, and seeing the size of some of the Glandinas it is quite easy to believe that these may number as many as a dozen in a day.

We have already mentioned the land shells, more nearly related than our true snails to the whelks and limpets of the sea, having abandoned the water more recently. Their curious snouts recall their

and the growth of the teeth of the rasp into long structures which are barbed like fish-hooks. The whole arrangement can actually be protruded for some distance from the slug's mouth. When this is done suddenly, it has a very striking, not to say startling, effect upon anyone who handles one of the creatures and is not aware of its capabilities.

It is when the animals are feeding, of course, that the apparatus is put to practical use.

We may imagine one of them crawling along in the burrow of an earth-worm, and if here it meets and touches the owner, both animals will recoil, and the contracted body of the mollusc will effectively block the passage. As nothing moves forward to meet the worm again, it becomes reassured and proceeds as far as the soft front of the slug, in which there may be a slight hollow. In trying to make its way through the obstacle, the pointed head-end of the worm may find its way, as Nature intends, into the depression and so tickle the slug. If this continues, the stimulation will cause the slug to open its mouth into which the depression leads and shoot out its worm catcher—the muscle already described. It is spoon-shaped at the end and the edges, which bristle with barbs, close inwards as the muscle comes to its full extension, so that not only does the anterior of the worm often get caught between them, but it is also impaled on the points, from which it cannot free itself owing to the barbs.

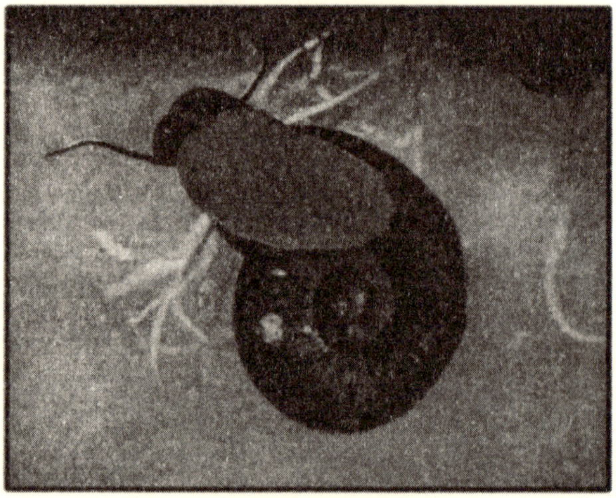

W. S. Berridge

WATER SNAIL FOUND IN POND AND DITCH

Of the water snails of Britain an important sub-family is that bearing the name Planorbis (flat coiled), and of this the Planorbis corneus is the largest. We see specimens here among some water weeds (top), and crawling up the side of an aquarium (bottom).

F. W. Bond

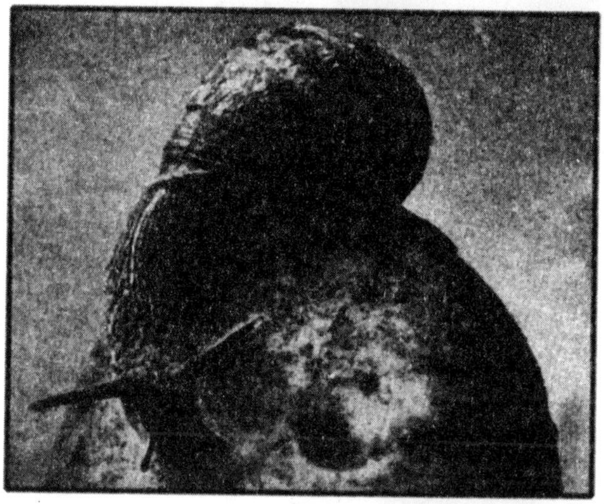

J. J. Ward

W. S. Berridge

NORMAL AND FREAK COILS OF THE GARDEN SNAIL'S SHELL

Below we see the normal form of the shell of the garden snail. Minor variations of banding and colour are common and, very occasionally, there may be an extraordinary freakishness in the actual coiling of the shell as it grows. The upper photograph shows an example of the ordinary garden snail like the one in the lower photograph, but which has had its spiral drawn out in a most astonishing way by some strange deviation from normal growth. The photograph shows the creature considerably enlarged.

gill-breathing relatives and forbears, although they are now quite adapted to a life in air.

There is an example found in chalky districts in this country not very unlike a small periwinkle, and its method of progression is very quaint. Instead of pushing forward the whole of its foot it moves each lateral half alternately, and crawls along as a man might make shift to walk if his ankles were not too tightly tied together.

INTERMEDIATE forms and those which can adapt themselves to varying circumstances have a special interest for students of nature, and the tropical apple snails are no exception. They still possess gills in addition to a lung chamber, and they enjoy an aquatic life, taking oxygen from the water or directly from the air through a tube, while in the dry season they can live on land. Their shells can at all times be tightly closed with a lid which is attached to their bodies and they seem to take advantage of any moisture that they can get. An explorer who kindly brought me several from places some two thousand miles inland in South America stated on his honour as an officer that they were discovered high up in trees in forks where the heavy dews that are common in that part of the world might trickle down to them.

The snails which we have been describing have two tentacles, and the eyes are not borne on the tips of these but are seated at the bases. We should also expect to find the same state of affairs in the fresh-water snails with doors to their shells, and not only is this the case with them, but also the air-breathing water-snails which have open mouths to their shells and come up to the surface for their supply are built in the same way.

Neville Kingston

LONG NECKS IN THE BIRD WORLD: SOUTH AMERICAN DARTER AND AUSTRALIAN SWAN

With the birds there is always a well-developed neck, even though this development is often not seen to advantage except in the skeleton. It is in birds that spend much of their time, or seek their food, in the water that we find the majority of long necks. The South American darter (left) for instance, seems rather overbalanced by its comparatively enormous neck when on land; but in the water the balance is seen to be perfect. Notice the curious angle or kink in the neck which seems to act like a spring and give an extra dash to the swift stroke of the beak when the bird makes the "dart" at its prey which has given it its name. On the right is a black Australian swan which holds its neck rather straighter than the European variety.

Chapter LXV

Long Necks and Short Necks

By Ruth C. Bisbee

Lecturer in Zoology, Liverpool University

A NECK is simply a narrow region of the body connecting the head with the trunk. A great many animals possess a neck in this sense, as, for example, bees and wasps, moths, butterflies and ants. In fact, all insects may be said to have both a neck and a waist for in all the body is divided into three regions—head, thorax, and abdomen, with constrictions in between.

In some insects, such as beetles, these constrictions are barely visible, and no casual observer would dream of speaking of either a neck or a waist. In the ants, on the other hand, both neck and waist are so excessively constricted that the creature appears to be in imminent danger of breaking into three pieces. Some of the more lowly creatures also have a neck in this sense of a mere constriction between head and trunk. The cuttle-fish and the octopus have clearly defined necks, as also have some of the water-fleas. The most vivid imagination, however, could not supply a shore crab with a neck.

The neck is best and most consistently developed amongst the backboned animals, and for this reason it is usual to think of a true neck as the neck of a vertebrate and to associate with it a fairly definite structure. The backbone together with the skull forms the axis of the whole skeleton, and naturally part of the backbone must pass through the neck. The backbone is divided into many small ring-like bones, the vertebrae, and these are specialised in structural detail in different regions of the body. Those in the neck, called cervical vertebrae, vary in number in different animals. The backbone is really a protection for the spinal cord, that part of the central nervous system which passes from the brain down the whole length of the body. The vertebrae are threaded on the spinal cord very much as beads are threaded on a string. Naturally the spinal cord passes through the neck, and it is the damage done to this cord when the neck is broken or dislocated which makes such an accident usually fatal. The food tube passing from the mouth to the stomach and the air tube passing from the mouth to the lungs also pass through the neck, as do the blood vessels of the head.

W. S. Berridge

SHAGGY-NECKED ALPACA

Belonging to the same family as the camel, but confined to South America, the alpaca has the characteristic long neck of its relatives. But a more erect pose is kept, the neck not sagging as in the camel's case.

All backboned animals possess most of these structures, but the region containing them is not always narrowed to form a neck.

In fishes the body is very rarely narrowed behind the head. There are one or two cases where there is a slight constriction, as in the sea-horse, but this is unusual. The other extreme is found in the sunfishes and anglers, where the body is actually wider just behind the head than in any other region.

In the amphibia—frogs, toads, newts and salamanders—the neck is developed, but it is very short and it is not always possible to distinguish it, externally, from the rest of the body. In some frogs and toads, for example, the head passes imperceptibly into the trunk.

Amongst the reptiles the neck is much better developed. Internally the cervical vertebrae are usually quite distinct from the rest, but vary much in number in the different forms. Externally, too, the neck is usually quite well defined, but there is great variety in general appearance. In crocodiles and in most lizards, for example, it is very short and thick, whereas in tortoises and turtles it is usually comparatively long and thin and flexible. In snakes it is impossible to say from external appearances where the neck really ends and the tail begins.

In birds the neck is always well developed, but varies tremendously in form. The cervical vertebrae are usually distinct from the rest but, as in reptiles, they vary considerably from one species to another. In the small birds of fields and woods, such as the sparrow and thrush, the starling and the robin, the neck appears to be very short and thick. When the feathers are removed, however, the neck is seen to be very much better developed than at first appears, although it is not exceedingly long. The condition is similar in most British common wild birds such as crows and rooks, hawks, owls, pigeons, cuckoos, etc., but there are some with much longer necks, as, for example, ducks and geese, swans, cormorants and herons. Outside the British Isles there is even greater variation, from the short-necked penguins to the long-necked storks and cranes and flamingoes.

Autotype

MASSIVE SHAGGY NECK OF THE AMERICAN BISON

The bison is remarkable for its ponderous shoulders, massive neck and strong, aggressive-looking head, and the more so as the hind quarters are much more slender than a mere view of the head and neck would suggest. The distinctive hump of the back is caused by the exceptionally long vertebrae joining the head to the body, and the neck itself is covered with shaggy hair.

The neck is often elaborately ornamented. There is a lizard in Australia which has a large frill of skin on each side of the neck, the two halves being joined together in front of the throat. This frill can be raised until it looks like a wide and elegant collar, or it can be lowered until it looks like a cape worn back to front. The cobras make themselves both handsome and terrifying in appearance by expanding the neck region into the famous hood. In many birds the neck is very beautifully decorated. In some the decoration consists simply of more brightly-coloured feathers than are found elsewhere on the body, as, for example, in male ducks. In others the feathers of the neck are specially long and loose, and in storks and herons and cranes they hang down loosely in front forming a kind of jabot. Others, again, wear a ruff as, for example the ruff itself, which has around the upper part of its neck a broad band of feathers which can be raised or lowered at will. In mammals, too, the neck is often ornamented. The lion's mane is a most handsome ornament and the quite different type of mane found in all members of the horse tribe is almost equally handsome and decorative.

Further, there is almost endless variety merely in the poise of the neck, especially in the way long necks are carried when their possessors walk or swim or fly, and in the way they are disposed of during rest. During flight, most birds stretch the neck straight out in front, and so perfect the stream-lines of the whole body. Ducks and geese and even flamingoes fly in this way. Herons, however, fly with the neck bent in the shape of an S, with the head drawn well back. Except during flight nearly all necks are bent, and most of them assume this S-shaped curve. In some one limb of the S is emphasised, in others another, but in nearly all the curvature is essentially the same. In the swan the S is evenly balanced and indeed the grace of the beautiful neck is proverbial.

In mammals the neck has a more definite structure than in any other creatures. Nearly all mammals have seven cervical vertebrae. The manatees have only six, and sloths may have six or eight or nine but, apart from a very few exceptions, in mammals the neck contains seven vertebrae. The form of the neck, however, varies greatly. There is the short thick neck of the whale which, from external appearances, would not be called a neck at all; the strong short neck of the elephant which is so admirably suited to carry the enormous weight of the head and tusks; the graceful curved neck of the horse; the ungainly-looking long neck of the camel, and the enormous tower-like neck of the giraffe. Yet in the giraffe and the whale alike there are seven neck vertebrae, neither more nor less.

Thus there is great variety in form in the neck region throughout the backboned animals, and the variation does not by any means end with differences in length and thickness.

In the giraffe the head and tail of the S are very short, the middle part very long and little curved, and the whole neck stands up awkwardly from the body like the arm of a mechanical crane. In tortoises and turtles the curvature of the neck is unusual. Most

Neville Kingston

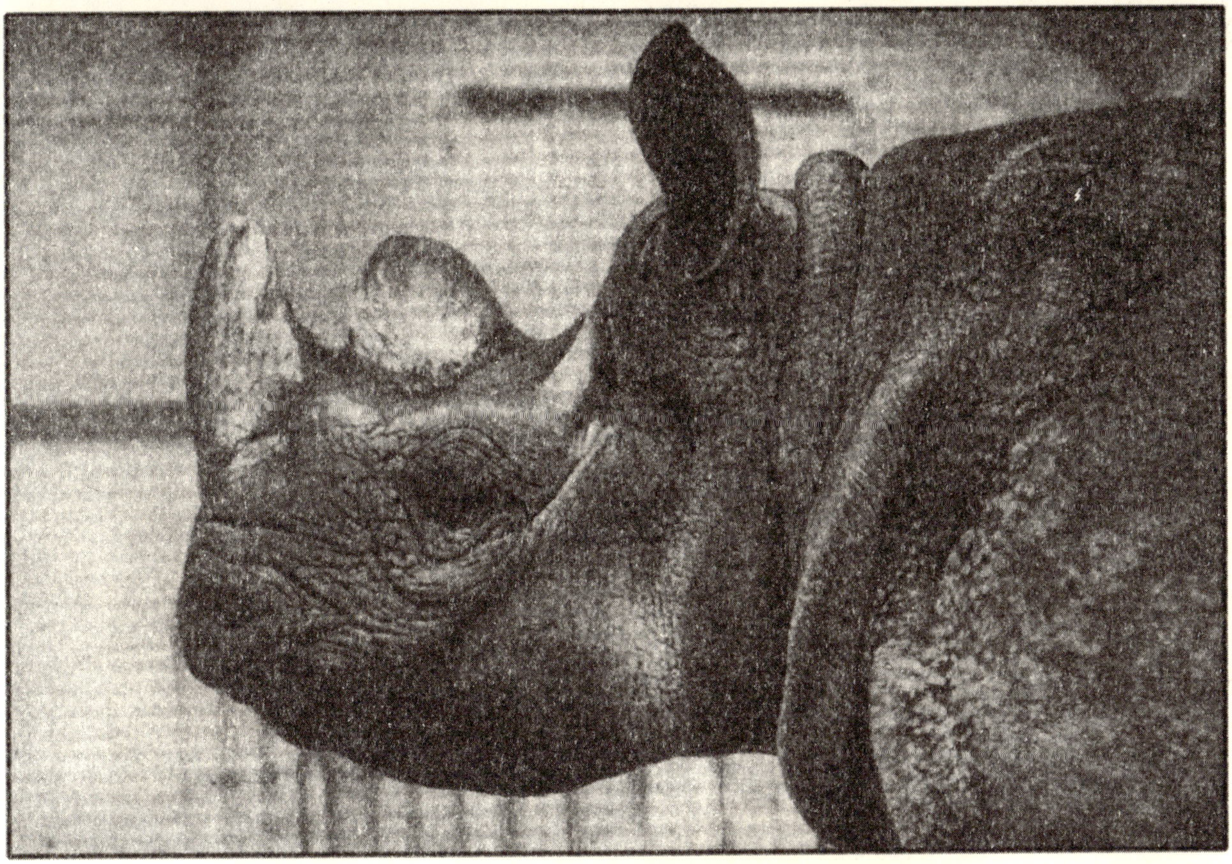

HEAVY-ARMOURED NECKS OF RHINOCEROS AND CROCODILE

Nature seems to have gone out of her way to make the neck of the rhinoceros invulnerable. Great folds of immensely tough hide envelop the neck, which is really longer than, at first sight, it seems to be. The heavy head tends rather to droop slightly when the animal is standing at rest. The crocodile (top), equipped with scale armour, is just as strongly protected. In fact, the bulge of the neck partly envelops the great jaws. The crocodile's neck is the handle of one of the most dreadful weapons owned by any animal.

EXTREME LIMITS OF THE "REACH" OF A GIRAFFE'S LONG AND TAPERING NECK

The giraffe is almost exclusively adapted for feeding on vegetation which would be out of reach of any other animal save, perhaps, the elephant which has the masterful habit of pulling down a tree whose leaves are too high for it to reach. What the giraffe can do when it tries is admirably illustrated here (left). The tree has been guarded by railings, but the giraffe easily overtops them and browses contentedly on branches from seventeen to twenty feet above the ground. But when it is a question of picking up food instead of reaching for it, or of obtaining some water, the neck, long as it is, is not quite long enough, and so the animal has to splay its forelegs as we see here (right).

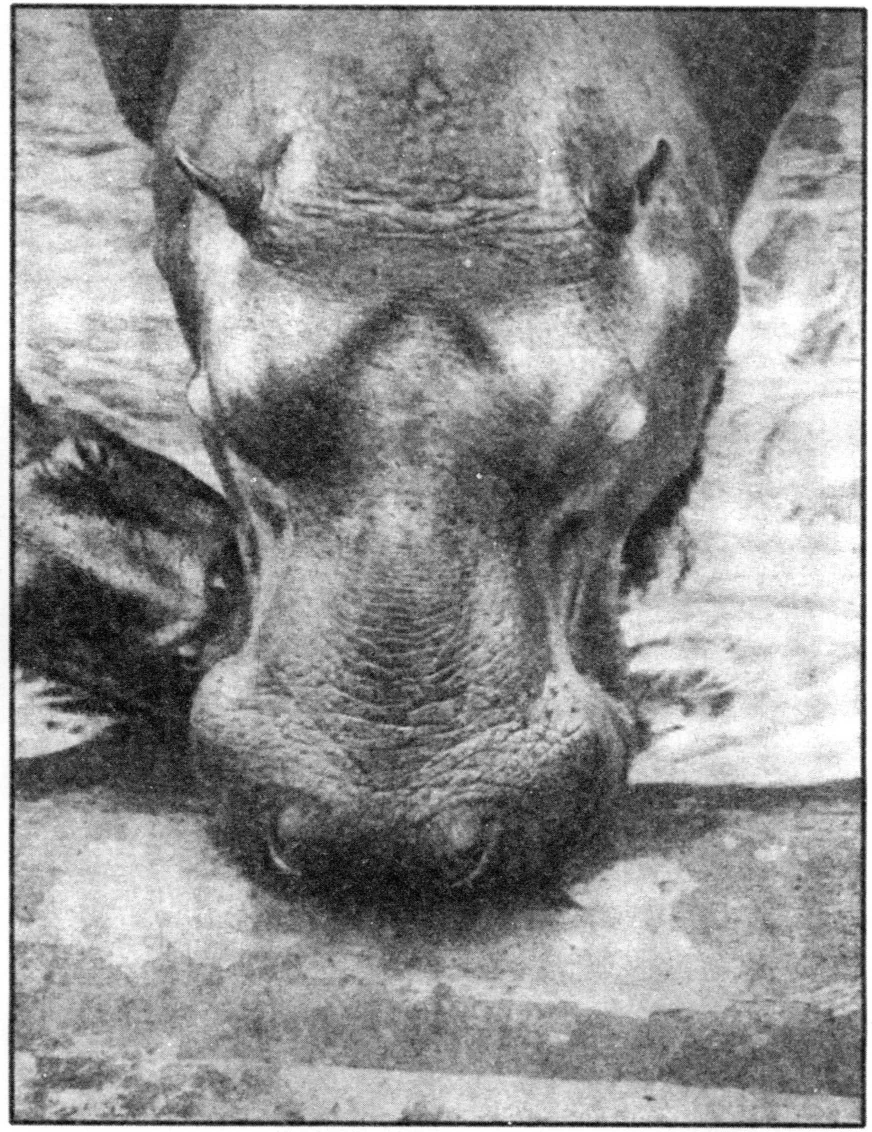

James

THE LONG AND THE SHORT OF NECK DEVELOPMENT: GIRAFFE AND HIPPOPOTAMUS

From the structural point of view, the neck is the connexion between body and head and its support is the spine, made up of sections of bone called vertebrae. The vertebrae of the neck are called cervical, and in most mammals there are seven cervical vertebrae. The whale, which has exteriorly an ill-defined neck, has seven and the giraffe, at the opposite end of the scale, has seven also, the difference in length being accounted for merely by the length of the individual vertebrae, or portions of the spinal column. The hippopotamus (right) has a neck thick enough to support its enormously heavy head and the neck is very short. Nevertheless, it has the same number of cervical vertebrae as the giraffe.

Neville Kingston

GIANT TORTOISE STRETCHES ITS NECK TO DRINK

Here we have an example of the disappearing neck. When the animal is alarmed the head is withdrawn and the neck is protected by the shell. But when a drink is needed the neck stretches to a surprising length. This is a giant tortoise from the Galapagos Islands, and the tortoises of those isolated rocks are exceedingly fond of water and of wallowing in mud.

region is completely immovable. In most birds, on the other hand, the neck is so flexible that it can be turned through fully half a circle. Most birds can look straight back over their tails and the majority sleep with the head turned back over the shoulder.

Thus, in this one region of the body, there is almost infinite variation, and one naturally seeks for some explanation, for some cause.

All long-necked animals are not related to each other; neither are all short-necked animals. Long necks are found amongst reptiles, birds and mammals. Short necks are also found in these same groups. But long necks are not found amongst fishes, nor amongst the amphibia. Can it be that short necks belong to the water, long necks to the land? But one remembers the short neck of the elephant and of the jerboa, of rats and mice and moles and shrews; and the enormously long necks of some of the ancient reptiles, the plesiosaurs, and the comparatively long necks of water-tortoises and turtles. Nevertheless, there does seem to be some connexion between habitat and length of neck.

Fishes form the only group of vertebrate animals whose members are all purely aquatic, and this is the only group where the neck is not developed at all. In the amphibia, whose members live partly on land and partly in water, the neck is developed but is always exceedingly short. Amongst reptiles the crocodiles live chiefly in water and their necks are exceedingly short; and amongst ancient extinct reptiles there were the ichthyosaurs, which lived entirely in water and were shaped almost exactly like a typical fish. But of course, as already mentioned, there are the water-tortoises and turtles and the plesiosaurs, all of them with long necks and all water-dwellers.

of these creatures have comparatively long necks, which are stretched forward as the animals walk or swim. During rest, however, the head and neck are usually partly or completely tucked away inside the shell. On these occasions the neck is bent in a peculiar way. In some it is bent in the form of a vertical S, but with the head at the bottom of the S. This is the case, for instance, in the typical land tortoises. In others, as, for example, the "Arrau" turtle of tropical South America, the neck bends horizontally instead of vertically, and the head is tucked away under the shell at one side.

There is also great variation in the flexibility of the neck. In some whales the cervical vertebrae are fused together into one continuous mass, so that the neck

AMONGST mammals the correlation between short, thick necks and life in the water is more striking. Whales are true mammals, breathing air and suckling their young, yet they are completely fish-like in shape with no sign of a neck externally. Internally, of course, there are the typical seven cervical vertebrae

W. S. Berridge

SURPRISING NECK OF THE TERRAPIN OF NORTH AMERICA

Terrapins are water tortoises and in North America, where they particularly frequent the salt marshes of the Atlantic coast, are very highly esteemed for the table. They are extremely active in the water and their long necks assist in the capture of fish, on which they prey, often with devastating local effects. When the neck is withdrawn into the shell (top) it is coiled in an "S"-shaped fashion, and no one, seeing the head just peeping nervously from between the upper and lower shells, would suspect its actual length (bottom).

excellent series of animals built essentially for life on land, but modified more and more for life in the water.

The advantage of the torpedo shape for a creature living and moving in the water is too obvious to require explanation. Man has yet to discover a more perfect shape for a body which is required to move quickly in this medium, and it is significant that so many creatures living in the water approximate to this shape. Whether the shape has developed as a result of life in the water, or whether water has been resorted to because the animals have varied in a direction which fits them for this habitat is a much debated question.

of the mammals, but these are very much shortened and flattened from back to front until they are like a series of plates. The dugong, another mammal which lives entirely in the water, also has an extremely short, thick neck and the body tapers posteriorly and is somewhat fish-like in shape, though much less so than that of the whales. Seals, too, are mammals which live most of their lives in water, and here again the body is rather torpedo-shaped with the neck strong, short and thick.

The powerful and forbidding-looking walrus spends much time in the water and is also, to some extent, torpedo-shaped. In fact, the walrus, seal, dugong and whale form an

A. H. Hall

POSTURES OF THE PELICAN'S NECK

While clumsy and, to human eyes, a little ludicrous on land, the pelican is very active in the water. The stretching power of the neck is well shown in the lower photograph where the long neck is enhanced by the length of the beak. The long beak (above) makes up for a long neck when breast feathers need preening.

Long and Short Necks

In regard to terrestrial animals it is also possible to see a connexion between the mode of life and the length of the neck. It seems to be a general rule that those creatures which live in vast, open spaces and have to travel great distances have long legs. There is the camel which travels across wide deserts; the ostrich which lives on the open, arid tracts of Africa, and the giraffes which live in the same regions—all with long legs suitable for covering great distances, whether in fleeing from enemies or in search of food. And if the legs are long, there must be some device to enable the creature to reach the ground with its mouth for food and water. In the elephant there is the trunk; in the curlew there is an extremely long bill; but in the vast majority of cases there is a long neck for the fulfilling of this purpose.

EVEN in the giraffe, which finds its food on the high branches of trees, it is still necessary that it should be able to put its mouth to the ground in order to drink. As a matter of fact, the giraffe's neck is barely long enough for this purpose, with the result that its characteristic attitude, when drinking, is extremely awkward. It places its front legs widely apart and stretches its neck down between them.

The same correlation between long legs and long necks is found amongst wading birds and is here strikingly associated with the method of obtaining food. Birds like storks, herons, and flamingoes live on aquatic creatures, but cannot dive for them. They wade out into deep water in search of food, and their long necks enable them to reach it.

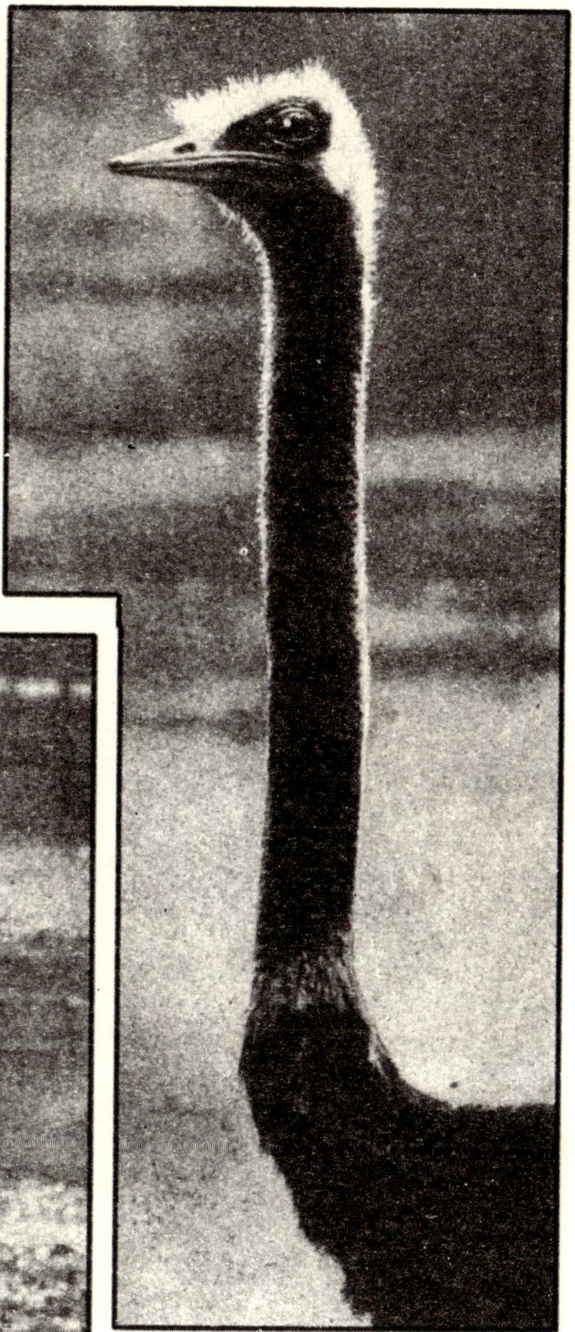

OSTRICH, A BIRD CONTORTIONIST

Long legs need a long neck to enable the head to reach the ground for feeding. But the ostrich's neck is capable of more than just enabling the beak to pick up food. It is even possible for it to be bent into a pot-hook at the bird's pleasure (bottom). Above is the ordinary and rather dignified carriage of beak and head.

James

FLAMINGO NECKS IN SLEEPING AND SWIMMING

One of the many delightful things about the flamingo is the eccentric but graceful pose which the bird takes up when resting (top). The neck has a kind of double twist in it as it lies like a snake along the back between the folded wings. When the bird is swimming the join of the neck and body is submerged (bottom), an unusual feature in swimming birds.

Nearly all animals with long legs have long necks, but by no means all animals with long necks have long legs. There were the ancient reptiles, the Plesiosaurs, already mentioned, which had enormously long necks, yet their limbs were short paddles. There are the swimming birds like ducks and geese and swans, pelicans and cormorants, all of which have long necks and short legs. These birds swim about in search of food and either dive for it like the cormorant or, like the swans, stretch down their long necks through the water and explore in the mud below.

Thus there is a very close correlation between the mode of

James

HERON NECKS ON THE ALERT AND WHEN RESTING

A peculiarity of the heron's neck is that it is tightly bent, during flight, into an " S," the head being held well back, whereas other long-necked birds stretch their necks while they are flying. A similar position is taken up when birds are at rest (top) though on alarm or when fishing the neck is quickly erected (bottom). These photographs show (bottom left) a cocoi heron from Para ; an egret, a white heron from the New World (bottom right) ; a nankeen night heron (top left), and a garden night heron (top right).

Long and Short Necks

James

CURIOUSLY SHAPED NECK OF THE STANLEY CRANE

Inhabiting the more open country in South Africa, but nowhere very common, the Stanley crane has an unusual neck which, gradually continuing the line of the body at the base, dwindles in the middle and then swells quickly to meet the curiously thickened head. The Stanley crane is carnivorous in its habits.

life and the shape and structure of the body, and nowhere is this correlation more striking than in regard to the neck. How this correlation has come about, however, we do not know, but two main theories have been put forward to explain it.

According to Darwin, evolution has taken place by a process of natural selection. He pointed out that no two living things are alike, and that many more creatures are produced than can be supported on the available food. He argued that there must, therefore, be a struggle for existence, and that in that struggle the fittest would survive, the unfit being exterminated. He took for granted that the survivors would hand on their own characteristics to succeeding generations, and that, in this way, distinct races would be evolved, each specially fitted for its own particular environment. The giraffe, for example, may conceivably have evolved in this way. Amongst its deer-like ancestors there would undoubtedly be insufficient food at some time or other, and there would then be competition. Amongst these creatures some would certainly be a little taller than others. With their longer legs they would be able to go farther afield in search of food than their shorter relatives, or with their longer necks they would be able to reach the shoots and leaves on the upper branches of trees, beyond the reach of their companions. These slight advantages might make the difference between life and death in times of great stress, and so the taller ones might live while the shorter ones died of starvation. If the taller ones handed on their tallness to their offspring, and if these again varied among themselves, some being even taller than their parents, and if the struggle continued, the advantage still remaining with the tall, gradually a very tall race might be evolved, such as the present-day giraffes.

LAMARCK, on the other hand, believed that evolution has taken place as a result of the striving of each individual to meet the demands of the environment. We know that use causes an organ to develop, and that disuse causes it to atrophy. Lamarck took for granted that the effects of use and disuse on the individual would be handed on through succeeding generations, and that in this way different races would be developed, each one perfectly fitted to meet the demands of its own environment. According to this theory the giraffe has evolved as a result of constant striving on the part of its ancestors to reach the fresher vegetation which is found near the tops of trees. Some of the deer-like creatures of the past quite likely did stretch upwards to reach fresh, juicy leaf-buds and young shoots. By so doing they would almost certainly increase their height to some slight extent. If this increased height were transmitted to their offspring, and if these in turn again stretched upwards for their food and so added a little more to their height, then gradually a taller race would evolve which might well give rise eventually to the giraffe of our own day.

Theoretically, then, both long necks and short necks could have evolved as a result either of natural selection or of the inheritance of the effects of use and disuse. There are, however, difficulties in the way of accepting either theory as a complete explanation.

TIGER BITTERN AT REST AND WHEN READY TO STRIKE

When the tiger bittern becomes excited the feathers on its neck, which are long, become fluffed out, so that the neck looks much thicker than it really is. This process leaves the back part of the neck uncovered save for a fine down. The thrust of a bittern's beak is like a dagger stroke, for the long neck is suddenly flung out as though a strong spring had been released. A wounded bittern is best left alone. The bird, celebrated for its booming note, is now very rare in Great Britain, though formerly plentiful.

W. S. Berridge

TRANSFORMATION OF MARKINGS FROM ADOLESCENT TO ADULT

If structural features have changed in any particular class of animals during the long process of evolution we can well imagine that the markings of the coat or hide have changed. It is believed that the young animal repeats, as it grows, the life story of its race. If this is so we must conclude that the Malay tapir (bottom) was once striped as its young now are. The very different markings of the adult animal are extremely striking. In the Indian wild swine (top) the adult's back is uniformly dark while the young have parti-coloured backs.

Chapter LXVI

The Wonder of Animal Transformation

By J. R. Ainsworth-Davis

Formerly Professor of Zoology, University of Wales

Among the wonder-tales of mythology, legend and fairy lore are many based on the belief that startling transformations of human beings into wolves or other animals were events of common occurrence. Such tales have always appealed to the imagination, and some of them will be remembered by every reader. They may be found in the "Metamorphoses" of Ovid ("metamorphoses" simply means "changes of form"), the "Arabian Nights," and many of the fairy tales that are always popular in the nursery.

Science has laid a ruthless hand on the beliefs of our forefathers, but she has made ample compensation by demonstrating that "truth is stranger than fiction," and the story of animal transformations is far more wonderful than any product of human imagination. By studying the life-histories of animals we shall find that many forms are startlingly unlike their parents when they make a start in life, and only attain adult characters by passing through a series of well-marked changes amounting to a complete metamorphosis.

A young animal strikingly dissimilar from the adult form when it begins life is called a "larva," the Latin word for ghost or spectre, the name being meant to suggest an obscure foreshadowing of what is to come. The most familiar examples are to be found in the insect world, and every one knows that the egg of a butterfly hatches out into a voracious larva called a caterpillar, this later on becoming a motionless chrysalis from which the adult insect (imago) ultimately emerges. The question "Why?" here naturally arises, and the answer partly explains not only this, but all other cases of metamorphosis.

The caterpillar is suited or adapted to conditions quite different from those which dominate the existence of the adult. It has to lead a special life of its own, and characters having relation to that life are dropped later on, new ones being acquired to meet the needs of the adult. The business of the caterpillar is to feed on vegetable substances that require chewing, and it possesses powerful biting jaws for the purpose. Wings are not wanted. The duty of the butterfly is to provide for continuance of the species, and wings help it in the search for a mate. The gross feeding habits of the caterpillar are abandoned, and biting jaws replaced by an elegant trunk or proboscis, used for sucking light refreshment consisting of the nectar provided by flowers. In the intermediate resting or chrysalis stage of life the body of the caterpillar is completely remodelled into that of the adult insect.

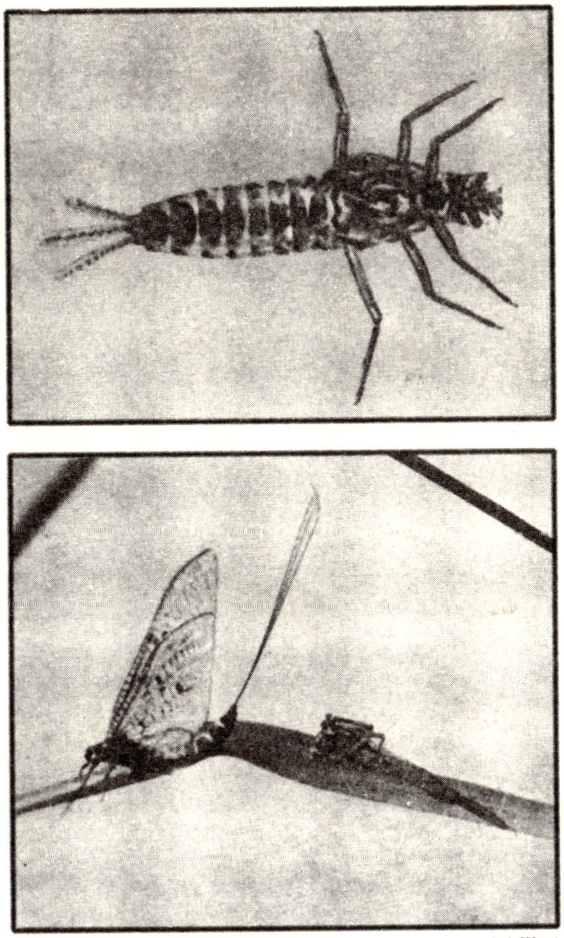

J. J. Ward

THE WONDERFUL MAYFLY

Although we know it as the most ephemeral of creatures, the mayfly's life is not so very short. Below the water in the larval stage (top) it spends many months and only emerges at the end to breed, equipped for that purpose and no other (bottom).

There is no larval stage in the highest groups of backboned animals—mammals, birds and reptiles—though in certain cases the young are born or hatched in a somewhat imperfect state. A baby, for example, is obviously a human being, so we do not call it a larva; nor do we apply that term to a helpless nestling bird. It is quite otherwise with the Amphibia, backboned creatures of which the best-known examples are newts, salamanders, frogs and toads, often erroneously called "reptiles." When grown up they possess soft and slimy skins, but have no scales and very rarely any claws.

The life-history of Amphibia is of particular interest, and proves that evolution is a possibility. No better text from which to preach Darwinism could be selected. The schoolboy who asserted that these creatures "die on the land and cannot live in the water" must have been ill-instructed in natural history, for we all know that they hatch out as aquatic tadpoles or pollywogs (larvae) which are ultimately transformed

It can therefore be concluded that the frog and its kindred have descended from fish-like forms, and it may be added that certain details in the structure of reptiles, birds and mammals can only be explained by a similar hypothesis.

The axolotl, a distant cousin of the frog, has a story to tell which is quite a romance. It is a newt-like creature with three pairs of delicate, feathery gills growing from the sides of the neck, and abounds in the lakes surrounding the city of Mexico. It lays eggs which hatch out into tadpoles that grow into axolotls, and from its first discovery by the Spanish conquerors of Mexico in the sixteenth century until 1865 it was not considered especially remarkable, except by the great French naturalist, Cuvier, who anticipated subsequent discoveries now to be described.

into land animals, though these are as a rule quite at home in the water as well. Let us take the frog as our most familiar native form of the kind. In spring the eggs (frog spawn) are laid in ponds, and the tadpoles hatch out in due course. They are adapted to life in water, and in many ways resemble fishes in structure, breathing dissolved air by means of gills, and being devoid of fore and hind limbs such as are present in the adult frog. As in a fish, the heart contains only impure blood, and pumps this to the gills for purification. After a time the tadpole is gradually transformed into a frog by a series of far-reaching processes of growth. The gills shrivel up, and as they do so a pair of lungs for breathing ordinary atmospheric air is developed. Fore and hind limbs grow out, the tail is absorbed, and the tadpole has now become a tiny frog which continues to grow until it attains its proper adult dimensions.

M. Duncan

SMALL SWIMMING CREATURE THAT TURNS INTO A LOBSTER

Lobsters, when first hatched, look very different from their parents. They escape from their eggs as swimming creatures which rise to near the surface, moving by means of the oar-like appendages on their legs The transformation is achieved by a series of moults, stage succeeding stage (top), till the perfect creature sinks down to a life in the depths (bottom).

IN the course of its life-history an animal is believed to repeat, in a general kind of way, the evolutionary stages passed through by its ancestors ; or, in other words, "climbs up its own genealogical tree."

Certain tailed amphibia, allied to newts but more addicted to terrestrial life, are known as salamanders, and a common European form was believed in the Middle Ages to be "bred and nourished by fire." The tiger salamander (*Amblystoma tigrinum*), so called from the yellow streaks and blotches on its dark skin, ranges from New York to California and Mexico, and until 1865 nobody suspected it had anything to do with the axolotl. But in that year some axolotls kept in Paris lost

ERRANT FINS OF THE FANTAIL FISH TRANSFORMED FROM THE NORMAL

By means of careful interference with the course of Nature during centuries there have been produced, by breeders in China and Japan, finned grotesques such as this which exist solely to satisfy the aesthetic appreciations of those who brought such very special gold fish into being. The artificial multiplying of the fins produced a transformation from natural utility to a humanly evolved decorativeness.

their gills, and underwent other changes by which they were converted into tiger salamanders. Several ways of bringing about this transformation are known, and the axolotl is now regarded as a sort of permanent tadpole which has grown limbs and taken to egg-laying, thus cutting the adult stage out of its life-history. Who can say how many sorts of animal have done the same thing? Perhaps we ourselves are nothing but human axolotls!

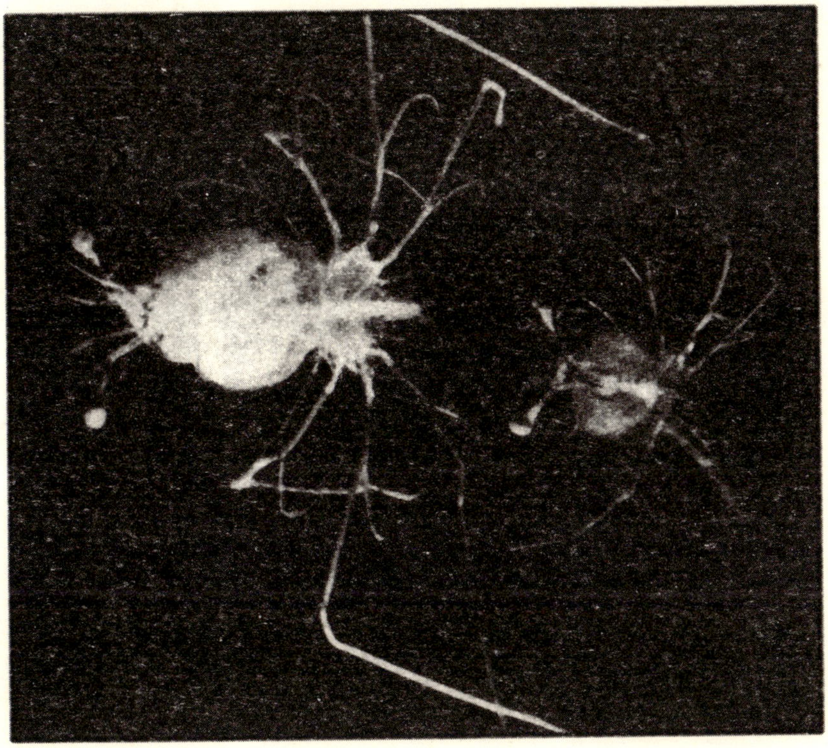

Many fishes hatch out as larvae or "fry," differing greatly in appearance from the adult, as in the case of the salmon, a marine form which ascends rivers to spawn. This species begins life as a tiny alevin, about one 25th of an inch long, with a little bag of nutritive material (yolk) projecting from the under side of its body. Before the adult characters are fully attained several further stages have to be passed through—parr, smolt, and grilse—and this takes several years to effect.

M. Duncan

CRAWFISH: ITS LARVAL AND ADULT STAGES

In their early stages crawfishes have a superficial resemblance to spiders. Their bodies are broad and very thin, and also have a transparent quality which, in the larger forms found swimming in tropical waters, has earned them the misleading name of "glass crabs." After several moults the spider or crab-like appearance is lost and the eventual physical characteristics are assumed.

The story of the common eel is far more remarkable, but was very imperfectly known until 1897. Even yet there are gaps to fill. The older naturalists were aware that spawning does not take place in fresh water, and suspected that eels migrated to the sea for that purpose, especially as vast numbers of tiny eels, or elvers, ascend our rivers in spring and early summer. This annual movement is known as the "eel fare," and is particularly noticeable in the Severn estuary, where at one time elvers were regarded as a particularly dainty food, either stewed or fried.

This is what happens. Male eels spend five and a half to six and a half years in fresh water, and female eels six and a half to eight and a half years, before attaining maturity, when they assume a silvery appearance. These "silver eels" descend our rivers to spawn in the Atlantic during winter and spring at depths of not less than five hundred fathoms. They probably die when the eggs have been laid, but at any rate do not return to their old homes in fresh water. It is believed that these eggs float and hatch out at the surface. However that may be, there is no doubt that a little, transparent fish-like creature, formerly classified as a distinct species (*Leptocephalus brevirostris*), is really a larval eel, for in 1897 two Italian naturalists kept specimens under observation and witnessed their transformation into elvers. (The strange story of the eel is told at greater length in Chapter V.)

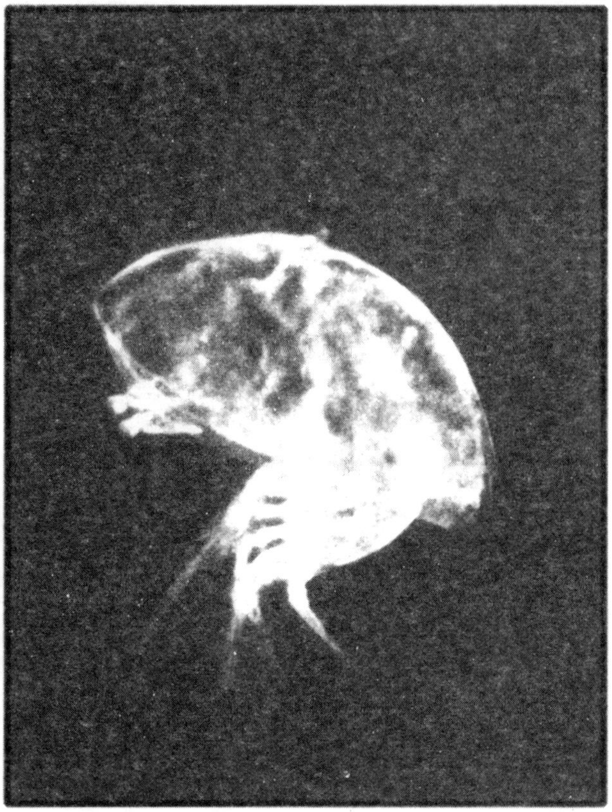

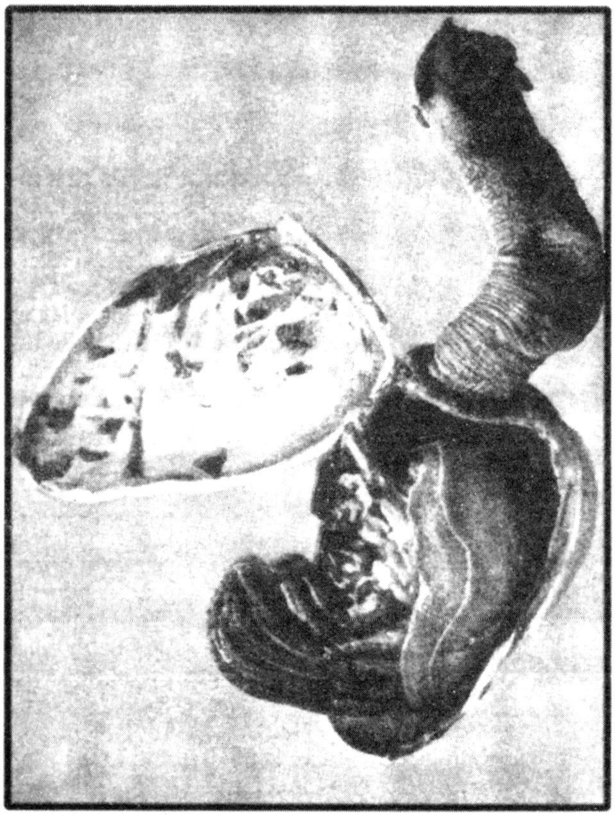

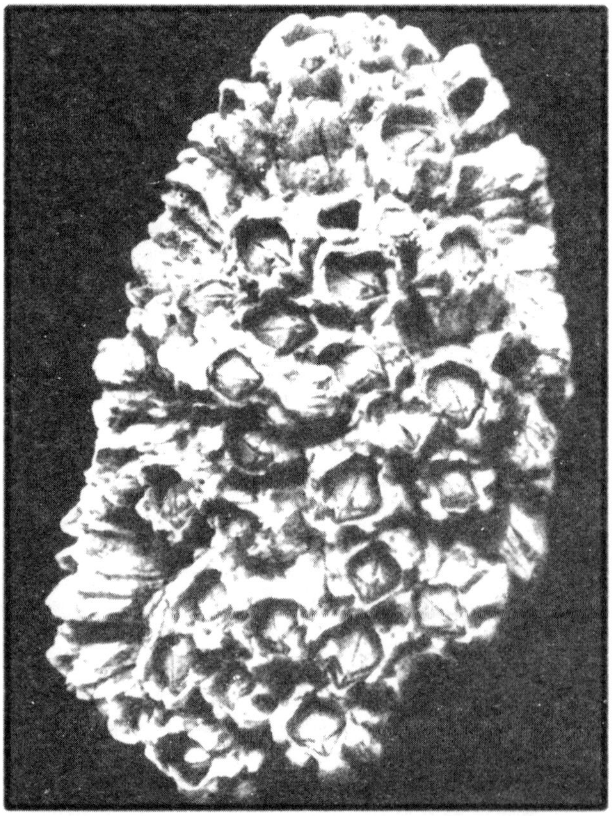

Martin Duncan

STRANGE LIFE STORY OF THE BARNACLE AS IT DEVELOPS

At the first the new hatched barnacle swims near the surface and then passes into a second larval stage in which it develops a shell and five pairs of legs, which it uses for swimming. It seeks and finds some rock or other solid object, and to this fastens itself. Gradually the entire head becomes firmly fastened and the long stalk is developed. These photographs show (bottom left) a ship's barnacle cut open to expose the internal organs; a group of acorn barnacles that live on rocks (bottom right), and (top) the early stages of the barnacle's life.

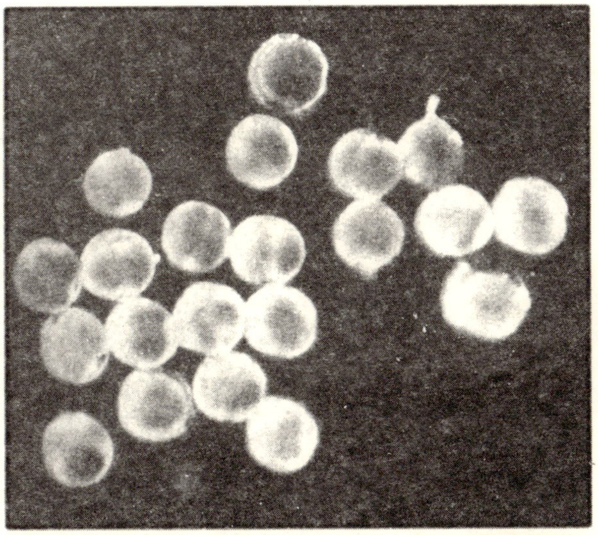

Eggs of the crab

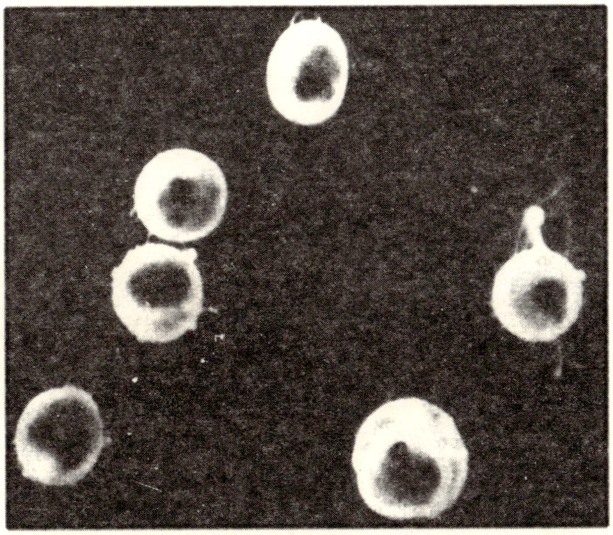

Larvae hatching from the eggs

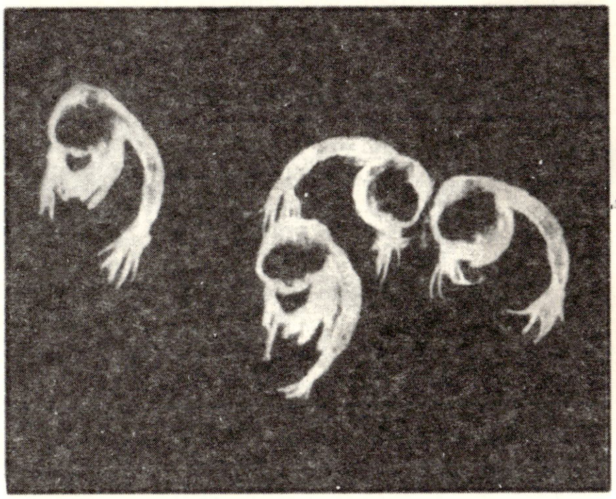

The larvae just emerged

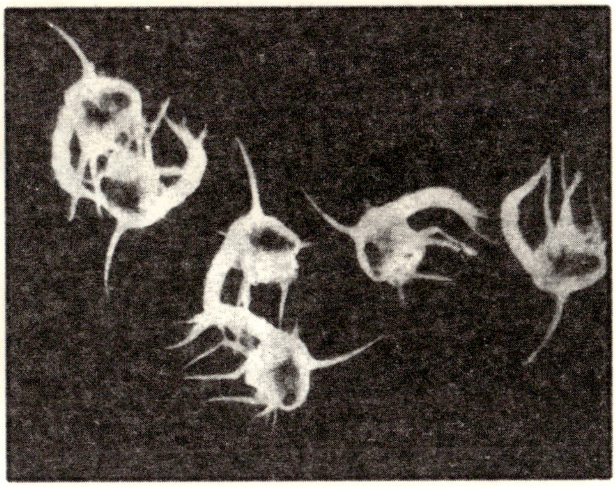

Early larval stage

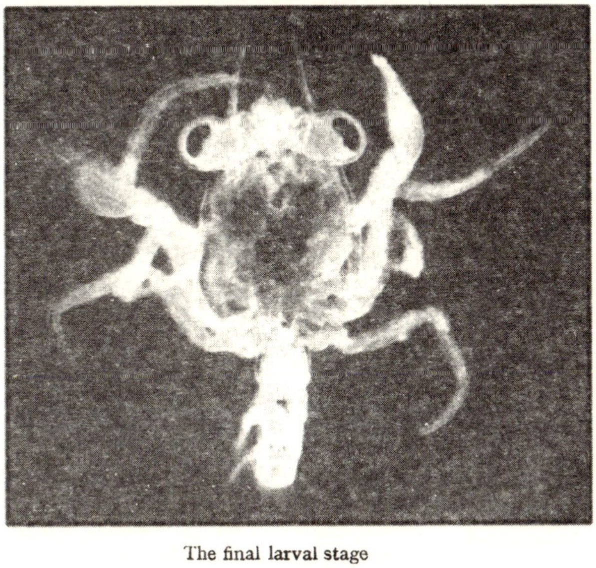

The final larval stage

Adult velvet fiddle crab

METAMORPHOSIS OF THE CRAB FROM LARVA TO FULL GROWTH

The shape of crabs does not differ between species and species nearly so much as does the form of each larva from that of its parent. Here we have the life-history of a velvet fiddler crab from egg to adult. The top left-hand photograph shows a crab's eggs just after fertilisation. When the eggs hatch (top right) weird shrimp-like creatures appear and swim actively about. This little larva gradually grows and develops till it passes through a lobster-like phase until finally the change is complete. Photos by F. M. Duncan.

AXOLOTL, A CREATURE OF MYSTERIOUS TRANSFORMATION

One of the most remarkable transformations in all Nature is that of the axolotl which lives in Mexico. Here we see an advanced stage (bottom) in the changing from a water creature to a land creature. The centre photograph shows a white variety and above it is a normally tinted specimen. (See also p. 487, upper photograph.)

The familiar flat fishes, including sole, plaice, turbot, halibut, and so forth, undergo remarkable changes in early life to fit them for life on the sea floor. They must not be confused with the skates and rays, which are related to the sharks and flattened from above downwards, the mouth being on the pale under side. But in the sole and its kindred the flat sides are not upper and lower, but right and left. The side which faces upwards, towards the light, is dark in colour, while the other side is white. It is a good example of protective coloration for, when the fish remains at rest, the hues and markings of the turned-up side harmonise perfectly with the surroundings. In turbot and brill the right side of the body is white, while the opposite is true for sole, plaice, and halibut. In other words, a turbot lies down on its right side and a sole on its left. In the former both eyes are on the left side, while in the latter they are seen to be on the right side.

The life-history of the plaice will show us how this curious modification has come about. It hatches out from the floating egg as a larva not unlike the alevin of the salmon, and the eyes are in the normal position. The little creature swims about for some time in the ordinary way, but gradually becomes flattened from side to side, and begins to tilt over to the left. Later on the right side becomes pigmented and the left eye moves round the edge of the head to join its fellow. The meaning of this remarkable movement is obvious. As the adult plaice spends most of its life on the sea floor hunting for food, with the pale left side turned downwards, an eye on that side would be useless, or worse, for it would be particularly liable to injury. Hence the transfer to the side which faces upwards.

The sea squirts, or ascidians (from the Greek, *askŏs*, a wine-skin), marine animals little known except to professional naturalists, undergo extraordinary transformations in the course of their life-history, and their place in the animal kingdom remained obscure until this had been studied. One of the simplest kinds, common on our shores, resembles in shape one of the skin receptacles in which the old easterns kept their wine, and which are alluded to in the parable about new wine in old bottles. It is fixed to some firm object by one end.

and there are two holes at the other end, while the body is covered by a tough protective coat or tunic, owing to which these creatures are often known as tunicates. Currents of sea water, bearing food and oxygen, flow into one of the two apertures mentioned, while products of waste are ejected from the other. When touched they contract and give out a jet of water, hence the name " sea squirt."

Whether these belong to the back-boned animals (Vertebrata) or back-boneless ones (Invertebrata) was long debated. The former possess, in some stage of their lives, three essential characters: (1) a backbone or its equivalent, (2) a tubular central nervous system (brain and spinal cord) situated near the upper surface, and (3) gill-slits, or perforations in the sides of the throat having to do with breathing. The adult sea squirt possesses only the third of these characters, making it at the best a very doubtful vertebrate, but it hatches out as a larva resembling a minute tadpole and possessing all three characters. After swimming about for a short time the larva attaches itself by means of a number of suckers on its head to some firm object, sheds its tail, and gradually takes on the adult form. It may, therefore, be regarded as a degenerate vertebrate.

THE Crustacea, which include crabs, lobsters, prawns, shrimps, and a host of other forms, often undergo striking transformations in the course of their life-histories, as in the case of the edible crab (*Cancer pagurus*). This familiar creature is one of the highest members of its class, and is greatly specialised in adaptation to its mode of life, which is that of a creeping sea scavenger. Its remote relative the lobster possesses a powerful jointed tail, by the alternate bending and straightening of which it is able to swim backwards with great rapidity. But in a crab the tail is not a swimming organ and has dwindled to an insignificant appendage tucked up out of danger on the under side of the very broad body, which is invested by powerful armour. The crab hatches out as a " zoæa " larva, with a respectable though limbless tail, staring eyes, and large " foot jaws " which are employed as paddles. This minute creature gradually increases in size, grows a couple of balancing spines at its front end, and

Mondiale

W. S. Berridge

DOG FISH AND THE CASE WHENCE IT CAME

In the lower photograph is a handsome specimen of the spotted dog fish. Dog fish start life in a kind of case which contains the eggs laid by their parents. The upper photograph shows some of these cases and where the sunlight catches them through the water it can be seen that they are almost transparent.

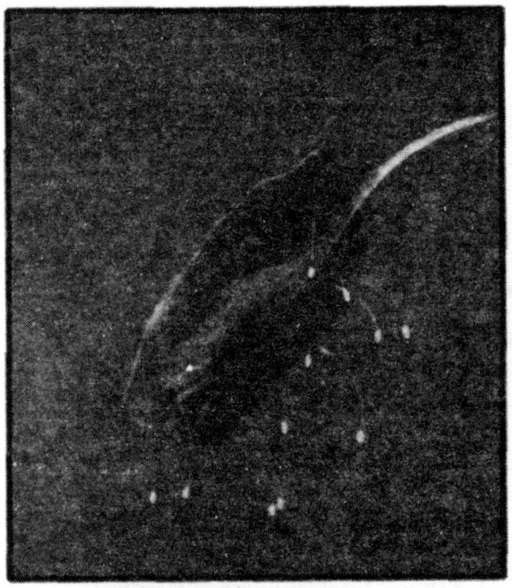

Lacewing fly's stalked eggs

Egg magnified

Larva magnified

Larva carrying skins of its aphid victims

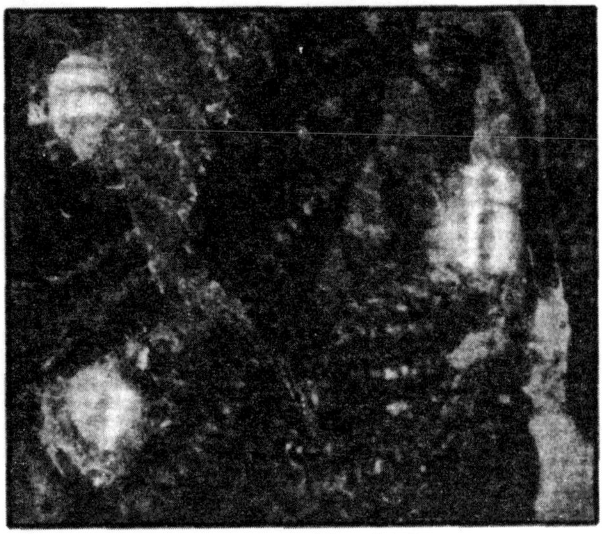

Silky cocoons of lacewing fly

Lacewing fly enlarged

Side view of lacewing fly, enlarged

J. J. Ward

FIERCE DESTROYER OF APHIDS THAT TURNS INTO A FLY

The microscope shows us some strange things, that look like hairs with the roots attached, to be the eggs of a lacewing fly (top left and top centre). From these emerge fierce-looking creatures with a tremendous appetite for greenfly. Above (centre left) we see one whose back is covered by the bodies of its victims. This creature eventually spins a cocoon, perhaps, as seen above (centre right), attached to bramble leaves, and emerges from its pupa case a perfect lacewing fly as shown in the bottom photographs (magnified).

Eggs of smooth newt with tadpoles inside

Young newts soon after hatching

Tadpole stage of newt showing gills

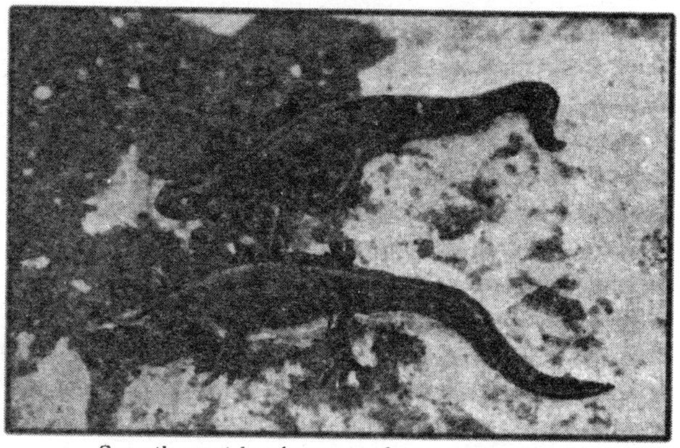

Smooth newt leaving water for terrestrial life

Newt depositing an egg which it folds in a leaf

Male and female smooth newts

CYCLE OF A NEWT'S EXISTENCE FROM WATER TO LAND AND BACK AGAIN

Newts, save in the breeding season (bottom right), are creatures of the land. The eggs are laid among the water-weeds, a leaf being folded (bottom left) to form a case. The eggs (top left) develop in a transparent envelope. When the eggs hatch the tadpole with feathery gills appears, breathing oxygen through the water. When development is complete and the gills have shrunk to be displaced by lungs, the newt leaves the water to live on land until the breeding season calls it back to water again. Photos by J. J. Ward.

W. S. Berridge

YOUNG RED DEER WITH SPOTTED COAT

In the world of backboned animals there is no startling change from one sort of creature to another, no metamorphosis so extraordinary that it would seem hardly possible for the last stage of the animal to have anything to do with the first. But there are changes of a minor sort, and these have a meaning. Compare this young deer with the adult in page 721.

acquires the full complement of limbs. Later on it passes into the lobster-like "megalopa" stage, which suggests that the remote ancestors of crabs were long-tailed animals, finally taking to a crawling life and acquiring the characters of the adult.

It may be mentioned here that the floating population of the sea, or plankton, includes a great many transparent larvae of all sorts and kinds, and it is probable that many marine animals have developed a larval stage in their life-history to facilitate distribution, by enabling them to settle down here and there in suitable places.

A crab is sometimes to be found with a round swelling on the under side of its tail, and dissection proves this to be a parasitic animal that extracts the juices from the unfortunate "host"—as the prey of a parasite is ironically called—by means of branching root-like threads. This parasite (*Sacculina*) is little more than a bag of eggs, which hatch out into larvae resembling those of the barnacles, undoubted Crustacea in spite of their appearance when adult, so that here we have to do with a sort of barnacle that has taken to a parasitic life and degenerated in consequence.

Among shellfish (Mollusca), sea snails and bivalves hatch out as minute free-swimming larvae extremely unlike the adults in appearance. We may take the oyster as an interesting case. It would hardly be suspected that this highly respectable but somewhat stolid animal had led an active life in its youth, yet this is the truth. It hatches out as an extremely small larva with a fringed flap on its head. The fringe consists of delicate living threads (cilia) which alternately bend and straighten, thus rowing the little creature about. The two valves of the shell develop on the right and left sides of the body. Later on, as "spawn" or "spat," the young oysters sink down to the sea floor and settle for life. Each of them lies down on its left side and becomes attached to some firm object by the left valve of the shell, this becoming concave for the reception of the body. The right valve remains flat and acts as a sort of lid The life-history of a fresh-water mussel (*Unio*, or *Anodonta*) presents several points of interest. There are two long flat gills, inner and outer, on each side of the body, and during the spawning season the outer ones become very much thickened, owing to the presence of innumerable developing larvae. Were these tiny creatures simply discharged into the surrounding water they would be carried away by the current, ultimately to perish in the sea. Such a fate is obviated by conversion of the outer gills into incubators, by means of which wonderful device the larvae do not escape until they are fully formed.

Each of them is then known as a glochidium, and is quite unlike the adult in appearance. It swims by

Autotype

ADULT RED DEER WHICH HAS LOST THE SPOTS OF ITS YOUTH

It is a very common thing amongst the higher animals for the young to differ from their parents in their exterior appearance. This difference takes the form of a striking variation in the markings and, in this case, we notice that while the young red deer, seen in the opposite page, has a spotted coat, its parent's coat is plain. This has suggested that, since the young animal is believed to recapitulate the story of the evolving of its particular species, the ancestors of the red deer were once spotted throughout their lives.

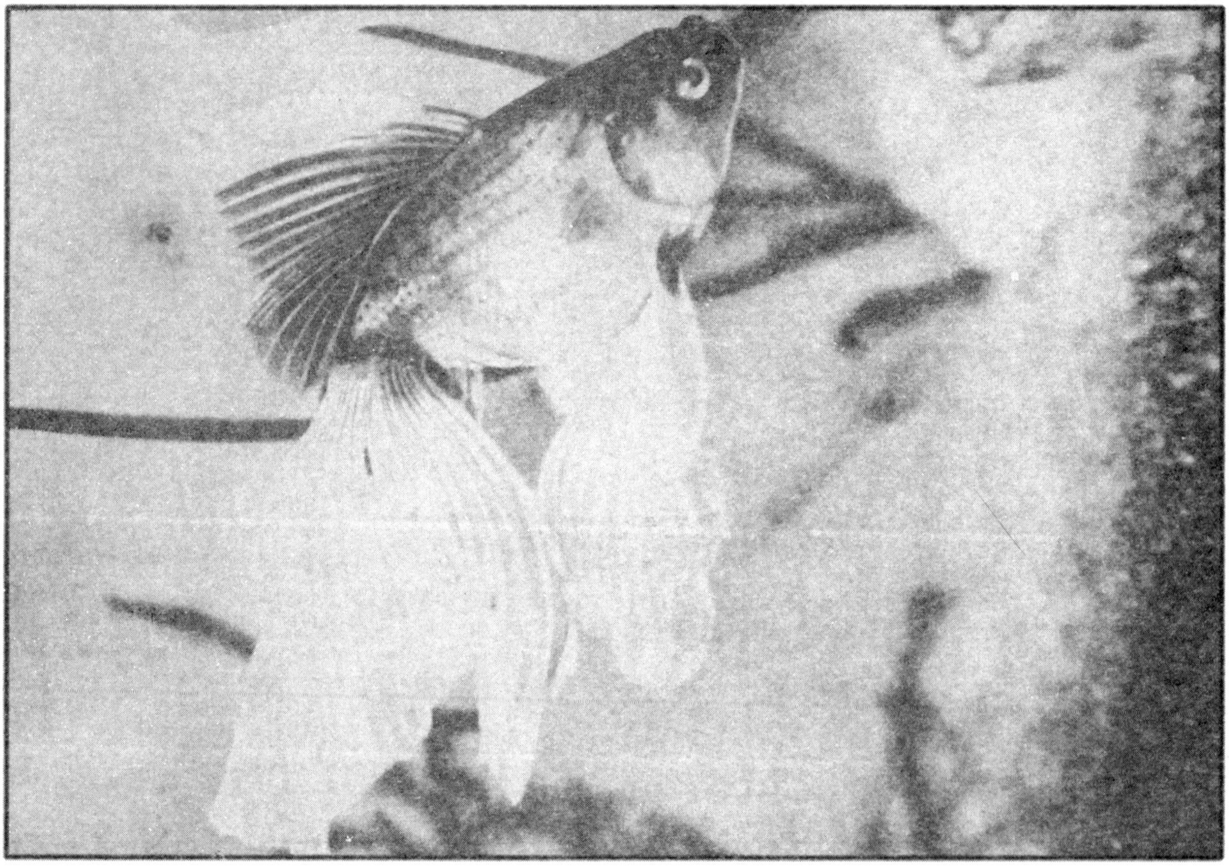

W. S. Pitt

CHANGES OF FORM WROUGHT BY MAN'S INTERVENTION

For centuries both the Chinese and the Japanese have been breeding goldfish, as is proved by their representation in old tapestries. More brightly or variously coloured specimens were sought for and changes of form were even brought about. The lower photograph shows a scaleless variety of the fantail goldfish, while above is a scaled fantail fish which, like all but the scaleless goldfish, starts life a bronze colour and becomes gold in a few months. The elaboration of fin, artificially wrought, is extraordinary.

FREAK TRANSFORMATIONS IN FOWL AND LIZARD

Among domestic fowls it is not uncommon for individuals to change their sex. Here is an example of a so-called "hen-cock" which, starting life as a hen, began to assume the plumage and comb of a cock. This is one of Nature's strangest mysteries. The upper photograph shows a specimen of the eyed lizard of Southern Europe. Having discarded its original tail, this one achieved the growth of two new ones, through some mysterious increase of virtue in the power which lizards have of reproducing lost tails

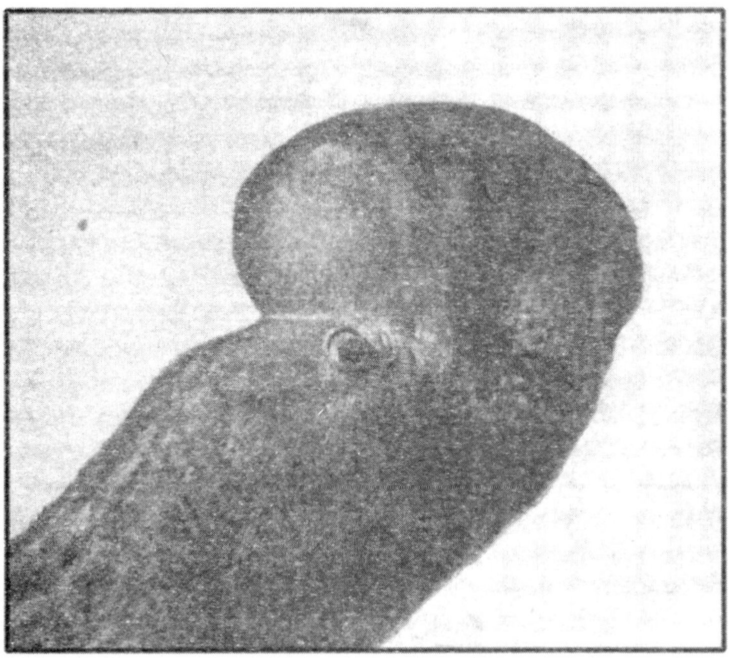

W S Berridge

ADULT AND YOUNG OF BLADDER-NOSED SEAL

Fiercest and most courageous of the seals of the Arctic Seas is the hooded or bladder-nosed seal, so called because of the remarkable bag of skin covering the nostrils of adult males. But the young, as we see here (top), are entirely without this distinguishing mark. The bag can be inflated with air.

alternately opening and closing the shell, and a long sticky thread growing from the under side of its body trails behind in the water. This holds fast should it come into contact with a stickleback or other small fish. The larva then grips the victim with its shell, each valve of which possesses a sharply bent hook. The irritated skin of the fish grows round the attached larva, forming a sort of gall. Here the glochidium is gradually transformed into a little mussel, the gall is ruptured by its growth, and it eventually falls down into the mud at the bottom, there to live for the rest of its life.

Hedgehog-skinned animals (Echinoderms), including starfishes, brittle stars, sea urchins, and sea cucumbers, usually hatch out as curious larvae which swim by the action of bands of cilia. That of a brittle star or sea urchin is called a pluteus, and looks something like a painter's easel turned upside down. The "legs," more numerous than in any easel, are strengthened by delicate calcareous rods. A series of complicated changes follow by which the adult form is ultimately attained.

Chapter LXVII

The Gamut of Sound in the Insect World

By L. C. Bushby

Curator of Insects, London Zoological Gardens

SOUNDS as well as scenes add to the enjoyment of a walk through woods and fields on a summer day or in the gloaming. There are the cries of animals, the songs and calls of birds, the sounds made by insects, and it is these last that to some minds have the greatest interest, inasmuch as close observation is needed to determine, first their origin and then the how and the why of both their mechanism and meaning.

Since the time when insects first began to be studied intimately by the old naturalists considerable attention has been paid to these problems and much knowledge has been gained on the subject; but, undoubtedly, further mysteries remain to be unravelled by dint of patient research.

Insects do not possess what can be termed a true voice; but many of them are able to produce by various methods sounds of one kind and another. All the main orders into which insects are divided contain species that have this power of sound production and it follows that as they differ in structure so they must of necessity employ different methods for the creation of their specific sounds.

The great majority of insect sounds are produced by the method known as stridulation, that is, the friction of one part of the body against another. The order Orthoptera, that contains, among other creatures, the grasshoppers, locusts and crickets, must be given first place in possessing the greatest number of expert exponents of stridulation. Who has not heard the monotonous chirping of the cricket and the more subdued rasping of the various grasshoppers? How are these sounds produced? Taking first the case of the crickets we find on investigation that it is from the wing-covers, or tegmina, that the sounds emanate. The tegmina are really modified fore-wings, and where they overlap each other at the base are provided, one with a finely-toothed area, and the other with a hardened process acting as a scraper. With a slight raising and lowering of the tegmina these two instruments are brought into contact with one another, producing a chirping sound that varies in shrillness according to species. There is in Africa a large cricket, *Brachytrypes membranaceus*, that constructs long burrows in the soil and at night sits at the mouth of the main burrow playing on its instrument. The noise it produces is exceedingly shrill and piercing, and can be heard nearly a mile away. It is a very vibrant note, filling the air to such an extent that should one be quite near a cricket when it ceases to stridulate the sudden silence affects the ear in a most amazing way.

In the Locustidae, a family containing the long-horned grasshoppers, it is again the overlapping bases of the fore-wings that carry the musical instruments and are capable of emitting some very quaint sounds, some of which are not unpleasing except, perhaps, in their monotony. The Katydids of North America, of which there are several species, are much noted for and, indeed, have gained their name by, the very quaint noise they make at nightfall. It sounds very like "Katy did, she did!"

The true grasshoppers and the locusts (*Acridiidae*) use a somewhat different form of instrument. The inner side of a portion of each hind leg is furnished with a row of fine teeth, and when these are rubbed up and down a raised and hardened vein on the fore-wings a rasping noise is produced. The sounds made by these insects are usually not so loud as those made by the crickets and long-horned grasshoppers, and may be heard during the day. A few grasshoppers can stridulate while in flight and effect this by the friction of the hardened upper surface of the costa, or front margin, of the hind wings and the thickened veins on the underside of the fore-wings, the legs not being used at all.

Several species of beetles (Coleoptera) make sounds in a variety of ways, and in most cases it is the file and scraper system that is employed. In some species the instrument is contained in the thoracic segments, the pronotum rubbing against the mesonotum; or it may be that friction is set up between rasps on the wing-cases, or elytra, and ridges on the abdomen, or between folds in the wings and abdominal ridges. Some species of beetle larvae stridulate, and in the majority of instances some part of the legs are used for the purpose. The larva of Passalus has the third pair of legs especially

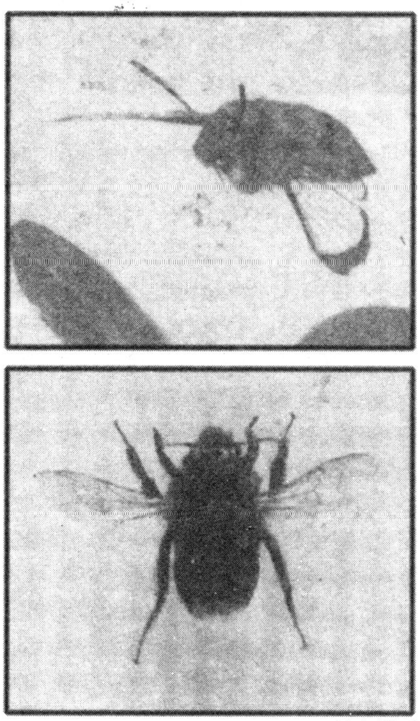

J. J. Ward

MUSICAL WINGS

Below is a humble-bee (worker) whose droning is so familiar a sound of summer-time. Above we have a bee-hawk moth which sips nectar while hovering on loud-humming wings.

Field crickets, male and female

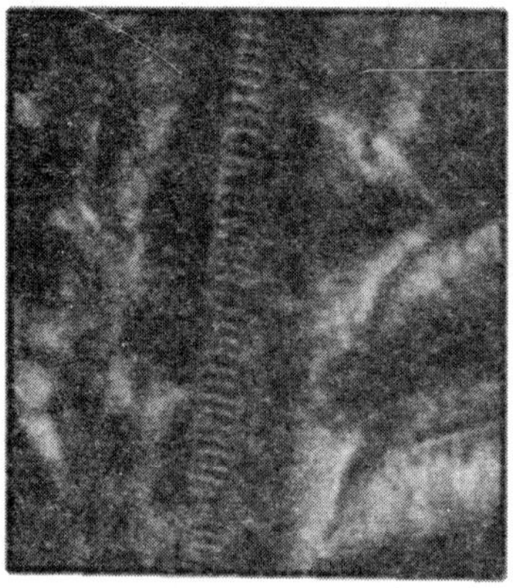

Stridulating bow from above

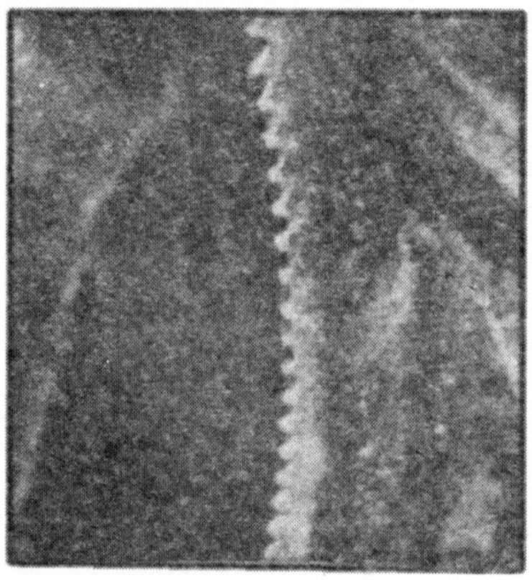

Side view of bow or file

Ear in foreleg

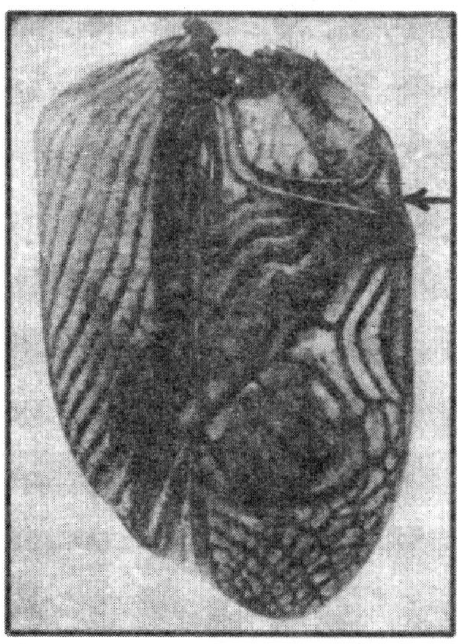

Bow on wing marked by arrow

CRICKETS AND HOW THEY CHIRRUP

Below are (left) the orifice of the cricket's "ear" which is situated just below the "knee" of the fore-leg: and (right) the fore wing-cover of a male field cricket, showing the file-like ridge or "bow" which is scraped on a vein of the wing beneath. The centre photographs give a general view of the "bow." The photographs are by H. Bastin.

modified and they are apparently used only for sound production. Many of the Scarabaeid larvae, while their legs are not so greatly modified, are able to emit squeaking sounds.

Certain butterflies and moths are capable of stridulation. The death's-head hawk-moth and other members of the same genus, *Acherontia*, emit a squeaking sound that is considered to be caused by air being forced through the proboscis and not actually by stridulation. It is said that the imago can sometimes be heard squeaking before it leaves the pupal case. The larvae of these moths can also make themselves heard; but in a different way; it is the mandibles vibrating sharply on each other that produce a sort of crackling sound. Clicking noises are made by a South American butterfly, *Angeronia*, when in flight, and are caused by a stridulating apparatus at the base of the fore-wings where there is a ridged and thickened area.

THERE are some species of ants possessing stridulating organs of one kind and another, although the sounds produced are in some cases not very audible. *Mutilla*, a species of parasitic wasp, can stridulate in much the same way as do the ants.

Wing-vibration plays a considerable part in the production of sound among insects and we have familiar examples of this in the bees, wasps, flies and certain beetles. The droning hum of the cockchafers or may-bugs (*Melolontha*) and the dor beetles (*Geotrupes*) as they fly at dusk is surely known to most people. The various bees and flies give out sounds of definite tone according to the rapidity of the wing-vibration, and the fact has been established that the wings of the honey-bee vibrate at the rate of 440 times in a second and produce the note A, while those of the house-fly vibrate at 335 and produce the note F. Some interesting experiments relating to wing-vibrations of bees and flies were carried out by Marey, who, with the aid of a revolving cylinder so arranged that it was just touched by the extremity of an insect's wing, was able to prove by the marks recorded on the cylinder that the strokes per second agreed very closely with the figures quoted.

How is it that a fly will buzz when held by the wings? It is the insect's breathing apparatus that enables it to protest so audibly. There is behind each of the thoracic spiracles a membraneous projection which vibrates during respiration and so causes the "voice." The sounds produced in this way differ in tone from those produced by wing-vibration and are usually higher in the scale. Bees and various diptera such as mosquitoes and hover-flies are all able to buzz without the aid of their wings. Most people will agree that the humming of a bee as it goes about its business in the garden is quite a pleasing sound, while a blowfly buzzing round a room only irritates. Why should this be so? It may be because of the actual difference in tone and, in the case of the blowfly, the jerkiness of it, or it may be that we associate the bee with pleasant things, flowers, honey, etc., and knowing the fly to be an intruder and a spoiler of our food we instinctively dislike his voice.

THE star songsters of the insect world are undoubtedly the cicadas. These insects occur in many parts of the world, the majority being found in tropical regions. They belong to the order Hemiptera and are provided with a proboscis with which they suck the juices of plants. They vary in size, some being less than an inch long and others robust creatures of three inches and more. There is one British species, but it is very local, living only in the New Forest.

Unlike small boys, who are supposed to be seen and not heard, cicadas are far more often heard than seen. Many of them frequent the high branches of trees in tropical forests, and it is only on hearing their loud, persistent notes that one is aware of their presence. The cicada does not stridulate; but possesses a vibrating instrument controlled by muscular action. This sound organ, the most highly specialised of any contained in insects, is situated at the base of the abdomen. It consists of a pair of drums, somewhat shell-like in shape. The membrane of these drums is in each case operated upon by a special muscle attached to the inner surface, causing them to vibrate and emit an intense shrilling sound. Two other membraneous areas, one of them known as the mirror, are situated near each drum and act as sounding boards, intensifying the sound. Only the males possess this remarkable apparatus, the

Side view of male cicada

Undersides of male and female cicadas

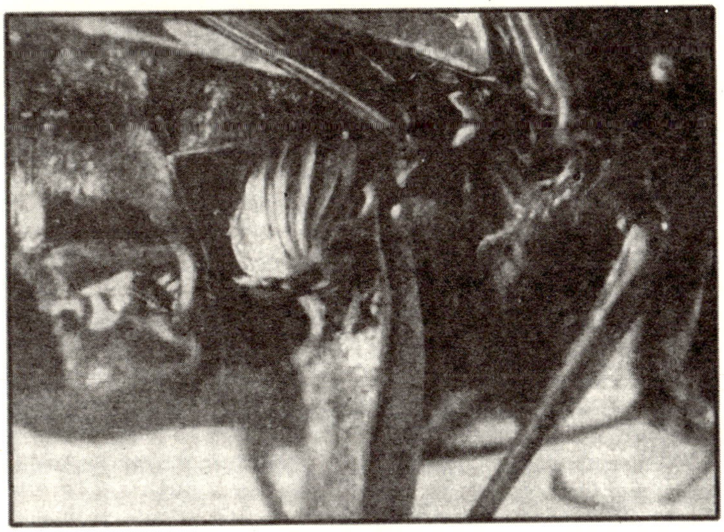

Drum of the male cicada

CHAMPION INSECT MUSIC MAKERS

Of all the insect "songsters" the cicada is the chief. Its sound organ or "drum" is the most highly-specialised apparatus of its kind in the insect world. The bottom photograph shows the drum at the base of the abdomen in the male. The top illustration shows a male on a twig. The photographs are by H. Bastin

Sound in the Insect World

females having, willy-nilly, to be silent. One other method of sound production must be mentioned. Its users require no specialised apparatus, but simply tap with some part of the body upon the surface on which they happen to be resting. Some beetles do this. A classic example is the death-watch beetle (*Xestobium rufovillosum*), a species that has gained much notoriety by damaging to a great extent the roof-timbers of old and famous buildings. Both sexes "call" by striking the wood on which they are standing with the lower part of the front of the head, the result being a series of distinct taps. Other insects that have this habit of tapping are certain species of termites, or white ants, as they are sometimes called, and it is the soldiers that perform by striking the flooring of their nest with the head.

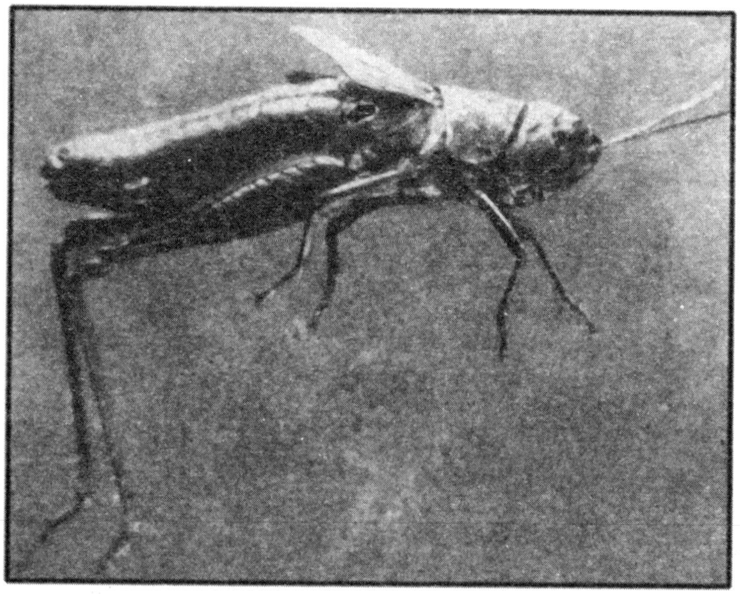

H. Bastin

GRASSHOPPER AND THE "FLYING GOOSEBERRY"

This young short-horned grasshopper (magnified) has its "ear" just below its wing-cover. Above is a species of locust called the "flying gooseberry," whose body is almost hollow and forms a resonator or sounding box when the insect rubs its hind-legs over certain ridges in its abdomen. The vital organs are in a very small area.

The natives of South Africa have given the name of "toktokkies" to certain Tenebrionid beetles that have the curious habit of raising the somewhat pointed end of the abdomen and letting it fall sharply, two or three times a second, on to the floor, thus producing a series of jarring taps.

WHY do insects make sounds? There must be some meaning attached to the fact that grasshoppers chirp and beetles squeak. Observations, both in the field and in the laboratory, have shown that there is in many cases definite reason for these audible outbursts. More often than not it is the males only that stridulate or in other ways declaim. This is the case with most of the long- and short-horned grasshoppers and the crickets. The male "sings," but the female is mute. They take the parts respectively of the serenader and the serenaded, for it is in the hope of attracting to himself the desired female that the cricket or the grasshopper performs his very best on his instrument. In the few instances where both sexes of grasshoppers possess stridulating organs, the female answers the male when he calls, her note frequently differing slightly from his.

The time most usually chosen for tuning up is at dusk or during the night; but quite a number of insects stridulate or hum during the sunny hours of the day, and it is reasonable to suppose that the sounds are then emitted for no other reason than mere enjoyment of life. In most species of stridulating beetles an equality of the sexes is the rule, the females being able to "answer back." There is an exception to this in the case of a Bostrichid beetle (*Phanopate*) of which the female has all the say and the male has to remain silent.

Death-watch beetles tap messages to one another through intervening pieces of timber, and appear to have established a definite code, for the series of taps they make always has the same rhythm. Fear or alarm is undoubtedly expressed by certain insects when they utilise their sound organs; certain beetles for instance, protest squeakingly when held or trapped in some way, and it is quite possible that the sudden, unexpected cry frightens their captors into dropping their prey so that it has a chance of escape. The death's-head hawk-moth is looked upon as a most uncanny creature by some people; the fact of its power of squeaking and having on its back marks resembling a skull and cross-bones have given rise to superstitions regarding it. It is believed that some insects sound the call of alarm on the approach of danger to warn others of the species. An excellent instance of this is known to occur in the nests of certain species of termites when a large number of the soldiers collect together and all tap the floor of the nest with their heads, maintaining some sort of rhythm apparently, while thus engaged. Some insects are able to perceive sounds. If this were not so sound production in these species would have no meaning. What kind of "ears" do insects possess? The grasshoppers and crickets, being the best musicians of the insect world, are provided with the most highly specialised form of auditory organs, and investigation has revealed the fact that these organs are situated either on the first pair of legs or on the abdomen.

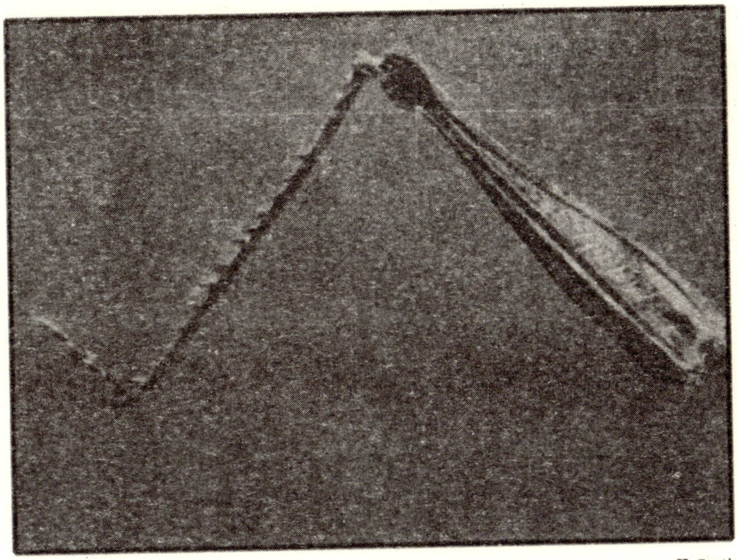

H. Bastin

SHORT-HORNED GRASSHOPPER AND ITS "BOW"
Magnification shows this creature's methods of producing sounds. Below is the ridge running along the thick or thigh part of the hind leg, and which is used as a bow to rub against a vein on the wing cover. Above is a general view of the whole leg.

It is in the Acridiidae, the short-horned grasshoppers, that the "ears" are carried on the abdomen. There is on each side of the base of the abdomen a membraneous area surrounded by a horny ring. This membrane, or tympanum, vibrates to certain sound waves and the vibrations are then transmitted to a nerve centre by means of specialised processes behind the tympanum. Many of the long-horned grasshoppers have "ears" on their fore-legs, a pair of auditory organs of a somewhat similar nature to those which have just been described being situated on each tibia. Crickets, also, hear with their fore-legs.

Can the humming of the female mosquito be heard by the male? The male mosquito is provided with beautifully plumed antennae and it is through the medium of these ornaments that he is able to perceive certain sounds. By experiment it has been found that some of the hairs respond to the vibrations of a tuning fork and that the vibrations are conveyed to the nerve centres by extremely delicate internal contrivances. In several other kinds of insects the organ of hearing is situated in the antennae. Some ants and bees, as well as the gnat-like flies (*Chironomidæ*) are able to hear thus and can be guided by sound to their waiting mates.

Besides the forms of auditory organs already mentioned there are others more or less specialised, some in the body-wall of the insect and others in the tibiae or tarsi of the legs.

Hearing is evidently not confined to the adult insects. In the larvae of some insects such as hover-flies, saw-flies, gad-flies, cockchafers, etc., there have been found, sometimes in the abdominal wall and sometimes on the legs, certain simplified but very interesting forms of auditory organs.

In this country we have a good many insects that make sounds of one kind or another. We hear the grasshoppers and the crickets reiterating their shrilling or chirping calls. In the flower-bedecked meadows and along the hedgerows we note with pleasure the humming of the bees as they go to and from the attractive blossoms. The vicinity of a lime tree in full bloom is most delightfully suggestive of some fairy concert hall

At dusk during the early summer we may hear the deep, droning hum of the may-bugs or perhaps a clumsy, heavily-flying dor beetle will pass just over our heads, sounding, as he approaches and disappears, like an aeroplane in miniature.

Sound in the Insect World

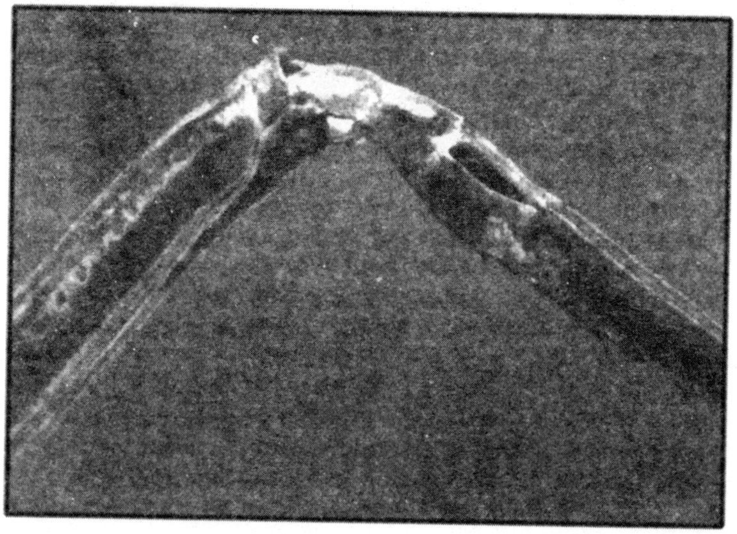

and some species create sounds that quaintly resemble certain words. The Katydids, of which mention has been already made, are celebrated for this. There are several species in North America and a few have gained great renown. The noise of a Brazilian insect, a long-horned grasshopper, is said to be very remarkable, a loud, resonant "ta-na-na, ta-na-na" being stridulated by the creature.

Tropical crickets are, many of them, very noisy, and it is difficult to believe that the extremely loud cries that some of the largest species make emanate from a mere insect. As with some of the cicadas, so certain of the crickets were, and still are, prized on account of their song. The Japanese confine them in little cages so that they may enjoy the music.

The sounds produced by British native insects are for the most part pleasing, especially when heard out of doors. We dislike the blue-bottle fly or the mosquito when they intrude into our houses and buzz in a very irritating fashion round the room or hum close to our ear as we lie in bed. But our insect musicians are few in numbers and weak in tone compared to those of tropical and sub-tropical countries. In practically all the books that have been written on the natural history of various tropical regions reference has been made to the predominance of insect sounds. In many cases the sounds have been described as being the very opposite of pleasant. The shrillness and monotony of some insect cries are said to be absolutely wearing to the nerves.

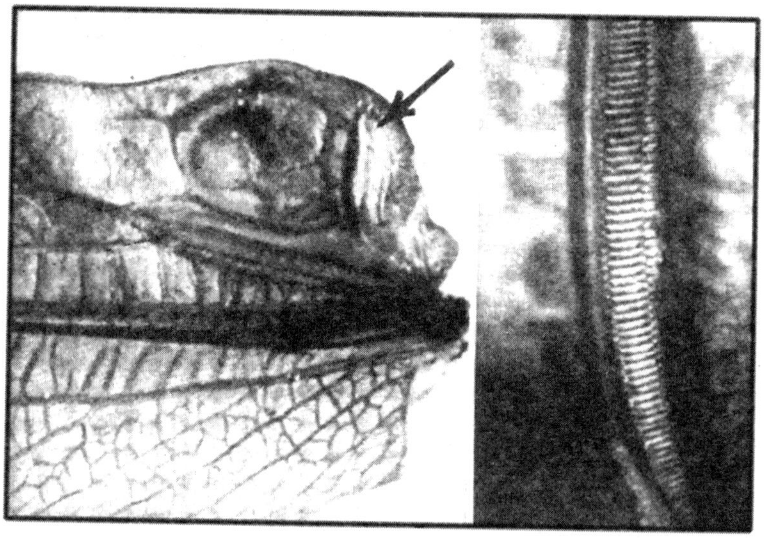

STRIDULATING VEINS OF THE GRASSHOPPERS — H. Bastin

These magnified photographs show (bottom left) the wing case of a large green grasshopper with the stridulating or grating vein, marked by an arrow, on which the insect rubs its legs to produce the chirrup; (bottom right) detail of the vein of a long-horned grasshopper and (top) the "ear" in the leg of a short-horned grasshopper.

It has been said previously that the cicadas are particularly noted for their musical powers. Travellers in jungle regions are often an unwilling audience at a cicada concert. Bates, in his book "A Naturalist on the Amazons," describes one of these concerts and says that the notes emitted were not only piercing and jarring at the start, but got shriller until they resembled nothing so much as a steam engine's whistle. It can well be imagined that the cicada can make itself heard a considerable distance away. Some of the lesser cicadas have more pleasing voices and, indeed, it is said that the ancient Greeks were in the habit of keeping specimens of a certain European species in cages so that they might appreciate their singing at leisure.

Many of the tropical grasshoppers and crickets have voices too loud and harsh to be pleasant and in others it is not so much the intensity as the monotony of the sound that irritates. There are some species, however, that are more or less pleasing to listen to.

It is unfortunate that the majority of the tropical insects choose the early part of the night in which to give vent to their vocal powers. Did they but perform during the sunny hours of the day they would doubtless be much more appreciated by the people who reside in or travel through the warmer climes and who are kept awake by the insects' chorus mingled with the cries of birds and monkeys.

It will be realized from what has been described in the foregoing that the main reason for sound production among insects is the mating impulse. Some insects, it is true, sing for no apparent reason; but with the majority it is a case of the male lifting up his voice in praise of, or in ardent appeal to, his loved one and she listening mutely or possibly intimating in faint tones that—she will.

Chapter LXVIII

Mammals That Live in the Sea

By R. J. Daniel

Lecturer in Oceanography, Liverpool University

It is part of our common experience that the domestic animals we see about us every day, such as the horse and the dog, differ very much from each other in appearance, but it takes a little reflection perhaps before we realize that in some important respects they are very similar to each other. They are, like ourselves, all warm blooded, they are all covered with hair, and the young are fed on milk and nursed with great affection.

Because of these general attributes such animals have been classed together as mammals, and the group shows a diversity of forms which have been successful in all parts of the world and under varying conditions of climate. So great has been this adaptability that certain of them have taken to the sea, and there they are able to prey upon such creatures as the fishes, which are themselves peculiarly fitted for life in the water, and the ancestors of which never knew any other domain.

The accommodation of certain mammals to a marine life has only come about gradually, and all have not become suited to the same degree. The carnivorous Pinnipedes, to which group the seals, the sea-lions, and the walruses belong, live mainly in the sea, but all may come ashore either upon ice or some quiet stretch of coast, and the young are never born in the water.

The second great order of marine mammals, the Cetacea, which includes the whales, have gone a step further and have taken to the sea so thoroughly that they have become totally incapacitated for any life on the land, and if they touch bottom, as may happen upon occasion, they become stranded like any ship and find themselves in a similar desperate position. The members of the whale family, in fact, have become so modified externally for dwelling in the sea that they are often wrongly referred to as fishes, a very natural error under the circumstances, and one that was made by students of Natural History for several centuries. Yet, in spite of these changes, which include the loss of practically all hair upon the body, the porpoises and whales are so much like the land mammals in general habits and internal structure that one does not hesitate in pronouncing them as being allied to this group.

The various members of the seal family themselves show degrees of liking for the water, and also represent different stages of agility when on the land. In all of them the limbs have been shaped into paddles which are peculiarly suited for aquatic locomotion, but these still possess a strong skeleton and are therefore able to assist also in land transport.

The walrus, or morse, of the polar regions is the least accommodated of this group when in the sea. While swimming it uses both its fore and hind limbs, but although powerful it is not so graceful in its movements as the other Pinnipedes. It rarely ventures out of shoal water, but moves about among the banks, seeking its under-water food, which consists mainly of clams. The male walrus, and many of the females, too, possess a pair of tusks in the upper jaw that are employed in digging for the shellfish. The very stout bristles that form the whiskers assist in this hunt and the grinding teeth crunch the clams that are procured.

W. S. Berridge

SEA-ELEPHANT OF KERGUELEN ISLAND

Sea-elephants, or elephant seals, are so called not only on account of their bulk—they are often twelve feet long—but also on account of the proboscis they possess, which rather resembles a very short trunk. Their chief home is Kerguelen Island in the south of the Indian Ocean

The walrus is very robust in appearance, this being due in part to the thick layer of blubber, or oil, that lies beneath its wrinkled skin. The animal lives a rough-and-tumble sort of life on the sea bottom or among the ice floes, so that it suffers numerous buffetings and lacerations. Many of the older ones, therefore, are scarred, and the covering of hair, which is never plentiful, is often worn away in patches. Altogether, the walrus is a tough and battered specimen, and the only member of the group that is really dangerous when hunted by man.

The eared seals, so called because they are the only Pinnipedes

DOLPHIN FOUND ROUND THE CAPE OF GOOD HOPE

Among the smaller species of dolphin is Heavyside's dolphin, named after the man who classified it, which haunts the waters round the Cape of Good Hope. Its average length is about four feet. The rounded back and the shortness of the "beak"—this beak being a characteristic of all the dolphins—make this particular species very reminiscent of the conventional dolphin of municipal statuary. In the colouring of its skin it rather resembles the killer-whale, or grampus. (See p. 456.)

that possess small external earflaps, are excellently represented by the sea-lions of the Southern Hemisphere and the North Pacific, and the sea-bears or fur seals of Alaska and the neighbouring islands of the Bering Straits. They are much more nimble on the land than the walrus; like that beast they are able to double the hind limbs under the body, and so progress with the assistance of the fore limbs, by means of a shuffling, rocking motion that carries them along a beach almost at a man's running pace, and, in addition, allows them to clamber up rocks in an amazing fashion. In such actions the animals are assisted by the flexibility of the torso, but this suppleness is as nothing when compared with the ease and rapidity of movement that takes place in the water. From the head thrust eagerly forward to the trailing hind legs used only for steering, the animal now presents a beautiful streamline body that is urged along with prodigious speed by means of powerful strokes of the paddle-like fore limbs. The eyes, which are somewhat short-sighted when gazing through the air, are able to focus themselves perfectly under water, so that in place of a soft and docile expression the sea-lion now displays the eager glinting eyes and the flashing teeth of the true hunting carnivore. Once in the sea it has few enemies it need fear, and even the ferocious sharks keep at a distance.

Eared seals that have been subject to a good deal of observation because of the value of their furs are the sea-bears, and their mode of life may be taken as typical of the group. They range the North Pacific in herds for seven or eight months of the year, hunting mainly for squids and different kinds of octopi that are in great abundance in the ocean.

Leaving the islands that may be regarded as their homes, they strike south to temperate seas then,

W. S. Berridge

COMMON DOLPHIN SOMETIMES SEEN OFF BRITAIN'S COAST

All over the Mediterranean and in many localities of the Atlantic one may meet the common dolphin sporting among the waves and following ships. It was this habit of swimming only half submerged that led to the famous legend of Arion, the Greek musician, who rode upon the dolphin's back, having charmed the animal by his playing of the lyre. The common dolphin is some seven feet long and has a dark back and a whitish belly. There are about fifty pairs of teeth to each jaw.

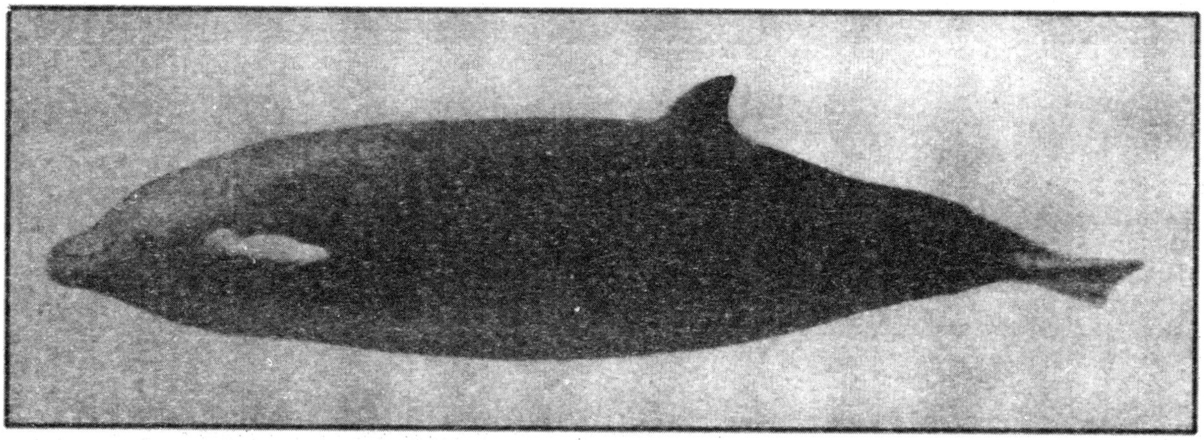

CUVIER'S WHALE WIDELY DISTRIBUTED THROUGH THE OCEANS

Belonging to the group of beaked whales in which the teeth are so vestigial that they never grow through the gum, Cuvier's whale is very widely distributed and has been found off the Shetland Islands and in the seas round New Zealand. The males of Cuvier's whale do possess two teeth but the females are toothless. These whales are less given to swimming in shoals than the dolphins and, for whales, are not very large, the average length being about nineteen or twenty feet. Their food consists for the most part of cuttlefish.

turning in towards the Californian coast, travel north again and return by a different route. During this lengthy course they thrive prodigiously, and so in the spring they approach the islands again in splendid condition. The adult males or "bulls" land first, and then follow the younger males and the females. The beaches are now swarming with them, and the reason for deserting the sea which has served them so well becomes apparent, for each mature female soon gives birth to a young one.

In spite of the noise and apparent confusion in these "rookeries," very definite social distinctions are drawn. Each of the older and stronger males gathers about himself a varying number of females and keeps jealous guard over them. There is a good deal of struggling and fighting, so that the younger males are cast out from these family arrangements and, deserting the rookery, they form a separate bachelor colony of their own. The young "pups," which are rather helpless at first, show no desire to enter the water, and are suckled by the mother. It is necessary for the latter, however, to hunt for food constantly, and so she takes to the sea again, and may leave the pup to its own devices for one or two days at a stretch.

It is only gradually that the young are persuaded into the water, and at first they are not able to swim with the rapidity and ease of the parents; they must become used to swimming as a child does to walking, and it is because of this that they are nurtured ashore. After three or four months on the beach, during which time the males have become very much reduced in condition, the sea journey is begun again; the pups are now able to accompany their parents, and when they return with them the following year they are still playful and ready to gambol with their younger relatives when these arrive.

W. S. Berridge

COMMON PORPOISE THAT OFTEN ASCENDS THE THAMES

The porpoise is one of the smaller members of its family and is usually about six feet long. The name is perhaps derived from the French "porc" pig, and "poisson" fish, and it is quite a regular visitor to the coasts of the British Isles. Specimens have been known to ascend the Thames as far as Mortlake on the Boat Race course. Notice that the tail is placed horizontally and not, as in fish, vertically, a distinguishing feature of all the mammals of the sea which belong to the great family of the Cetacea.

SCHOOL OF WHALES STRANDED AND HELPLESS IN A SHALLOW BAY NEAR CAPE TOWN

Many whales are gregarious and follow a leader just as sheep do, so that if one gets stranded by venturing too close to a shelving shore, the chances are that the whole "school" will share its fate. Our photograph shows a large school of more than one hundred whales which became stranded in a shallow bay not far from Cape Town. They lay about helplessly for some days, impotently lashing their huge tails, and some of the largest took nearly a week in dying. Meanwhile the local coloured population gathered at this wonderful chance of obtaining an unlimited supply of meat and cut steaks out of the unhappy creatures which were alleged to rival mutton in flavour

W. S. Berridge

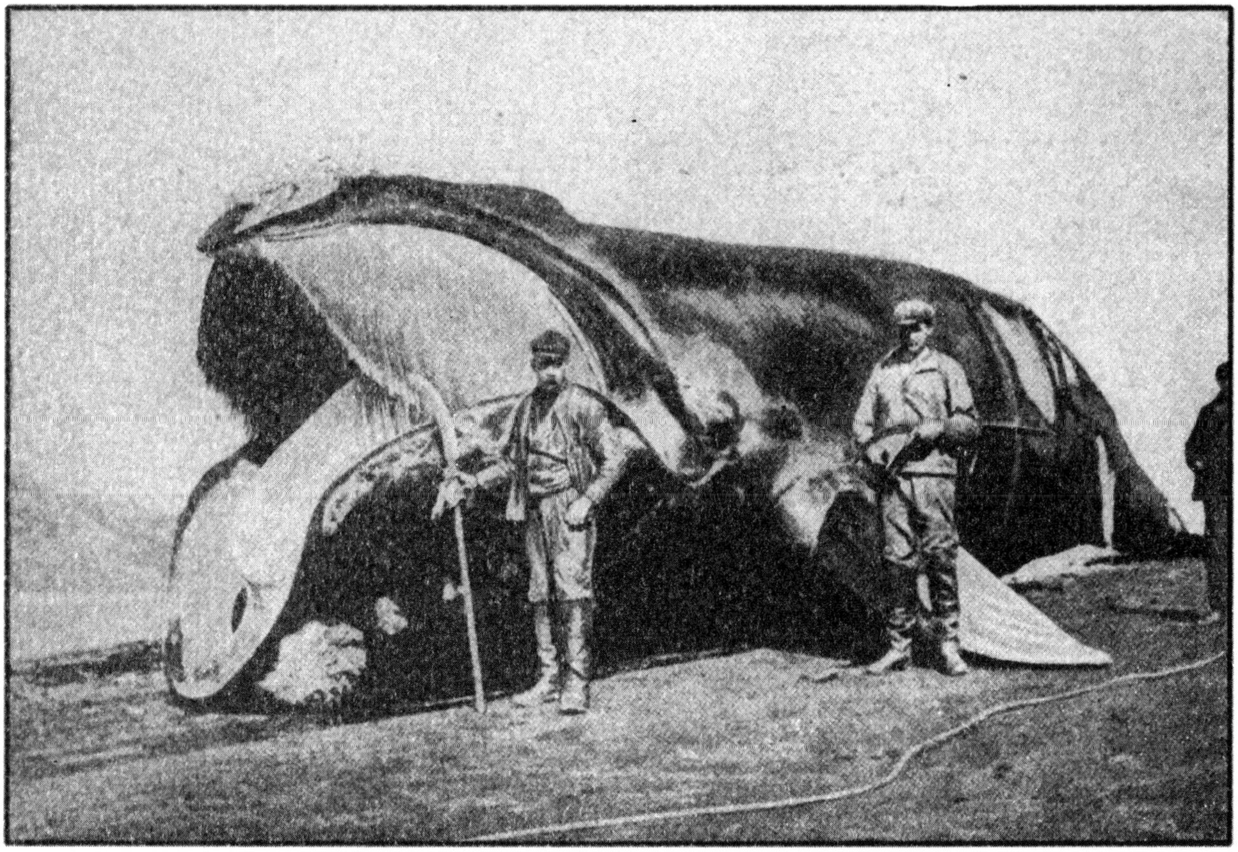

BALEEN WHALE AND A SCHOOL OF PILOT WHALES STRANDED IN MOUNT'S BAY

Baleen or whalebone whales, a specimen of which is shown in the lower photograph, are toothless but have evolved instead a kind of sieve inside their enormous mouths. With open jaws they rush through the water engulfing myriads of animalculae. At intervals they close their jaws and force the water through the "sieve" of whalebone which allows the water to pass out of the lips but retains the animalculae which are swallowed forthwith The upper photograph shows a school of pilot whales which has run ashore in Mount's Bay, Cornwall.

W. S. Berridge

DEAD RORQUAL ON THE ROCKS OF THE YORKSHIRE COAST

These photographs of a stranded whale were taken on the rocky foreshore at Cloughton, just north of Scarborough. Four species of rorqual are commonly found in the British waters, the one most frequently seen being the finner, which is from fifty to seventy feet long. The rorquals have a special modification of the throat and mouth, which are formed in a series of longitudinal pleats which allow of a considerable amount of stretching when the animals are feeding. The structure of a whale's skeleton makes it helpless when stranded.

MONSTER WHALES AT SOUTH GEORGIA AND SPITZBERGEN

The blue whale is the largest animal in the world and a specimen thirty yards long is on record. In the old days monsters like this were not encouraging to attack when whaling was carried on from the ship's boats and the harpoons were thrown by hand. Since steamships and harpoon guns were invented, however, even the blue whale can be "landed." Below we see two of these enormous creatures on the island of South Georgia. Above is a whale's carcass inflated to keep it afloat.

E.N.A.

ADVENTUROUS SEAL CAUGHT IN AN OLEARIA ON ENDERBY I.

Here we see a seal which, boldly adventuring inland, has become entangled in an olearia bush. This shrub, a native of certain parts of Australia, grows very close and thick. The photograph was taken on Enderby Island, one of the larger members of the Dampier Archipelago off the north-west of Australia. The seals are very fond of coming ashore to bask.

It is the true seals, however, that are most modified for aquatic life, and in consequence the most clumsy when ashore. Members of this group are almost universally spread about the coasts of the world, and excellent examples are found in the common and grey seals of our islands. The front limbs, or flippers are used in the water mainly for balance, and it is with the more powerful hind limbs that they swim; these are webbed, turned in towards each other and always point backwards from the body, so that they are of no great use on the land, and the seals must therefore heave themselves along in a laborious fashion. Nevertheless, they are fond of coming ashore in company and basking upon rocky ledges, on sand banks that are not very accessible to man, or, in the case of the Arctic species, upon the ice. The young are born out of the water and suckled for about eight weeks. In the case of the grey seal the mother when nursing only leaves the sea to suckle the young one. The movements then of both parent and pup are very awkward, and it is with some exertion that they drag themselves along to each other. Apparently the youngster may take to the water several days after birth and swimming seems instinctive to it, although naturally it plays about in the water close to the mother and may be assisted by her.

Belonging to the true seals also are the sea elephants, so called not only because of their bulk, for they may be twelve feet long, but also because the older males of the species have a short proboscis that is reminiscent of an elephant's trunk. They are common to the islands of the South Atlantic and the South Indian Oceans, but have suffered from the ravages of man, and are now centred on Kerguelen Island, where thousands of them may occupy the beach at one time. They are extremely cumbersome in their movements when ashore, but resort there for breeding purposes, during which time they have much the same habits as the eared seals. The females give birth about the end of September, and the migration to the Antarctic Circle begins in the following April.

So with the walrus, the eared seals and the true seals we have amphibious animals which may keep about the coast or make tremendous oceanic voyages, but which in all cases show indications of their land ancestry by resorting to the beach in order to give birth to their helpless young.

The conditions of life are very different with the whales and dolphins (or Cetaceans as they are called). Here also we have warm-blooded, air-breathing mammals, but they have become modified for aquatic life to an extraordinary degree. The smooth, oily skin is practically hairless, the fore limbs, or flippers are comparatively small, and the hind limbs, as such, have disappeared, although traces of the skeletal portions may be found within the body and,

SEA-LION IN ITS POOL AT THE ZOO

This gives us a good idea of the sea-lion's equipment for eating. We see one which has come to the edge of its swimming pool, in the London Zoological Gardens, in the hope of being fed. The wide-open jaws show the hungry eagerness that characterises these ravenous swimmers of the North Pacific and the seas of the Southern Hemisphere.

in addition, a strong muscular tail has been developed. Now, such modifications for life in the water are not unique, for they were adopted by certain of the large extinct reptiles which, being air breathers, also went back again to the sea; but they are drastic external changes nevertheless, and so disguise the fundamental mammalian nature of the Cetacea. The problem of reproduction in this latter group has been solved in a very different manner from that of the Pinnipedes. The young are perfectly formed within the mother long before they are born, and when brought forth in the sea they are comparatively large in size and able to swim alongside the parent; in the act of suckling the milk is forcibly pumped into them so that it may not by any chance become mixed with the sea water.

As already indicated, the motive power has been transferred from the limbs, which act mainly as balancing organs to the horizontal tail. When being used for slow swimming or turning in a confined space the flukes of the tail have a semi-rotary or sculling action, but speed is probably obtained by rapid upward and downward movements of that organ; at least, that is the impression given when looking downwards at porpoises playing about the bow of a ship.

When breaking surface a whale ejects the air from its lungs, or "blows" with great violence, and may breathe deeply several times before descending again.

Its body heat is about 96 degrees Centigrade, and a blanket of fat or blubber that lies immediately below a thin layer of skin prevents a rapid conduction of this heat to the surrounding water: the blubber is of a fibrous consistency, very resistant to strains, and yet elastic enough to allow freedom of movement for breathing. The smooth appearance of the head is accentuated by the absence of external ears, but nevertheless the hearing is probably acute to water-borne sounds. The eyes also are modified in structure in order to function in this medium; they are small, placed laterally and strongly made to withstand pressure, and are protected further by an oily substance which is exuded as tears. Most of the members of this group are migratory in their habits, and each kind moves about in "schools," being fond of company and affectionate in nature.

Kindred to the larger whales are the dolphins, which do not grow to any great size except in the case of the grampus or "killer." This beast may attain to twenty feet in length and like all dolphins it has a formidable array of teeth in both its top and bottom jaws. The "killers" are bold and ferocious in nature and fond of hunting in packs, gliding along at great speed with the black dorsal fin breaking the surface of the water. They range the oceans and are common in the waters of the Antarctic, where they prey upon seals among the ice. To this family

EXCITEMENTS OF CAPTURING AND DELIVERING SEA-ELEPHANTS FOR A ZOO

When the San Diego Zoological Society, of California, U.S.A., wanted some sea-elephants or elephant seals for their gardens, an expedition set out for Guadalupe Island, which is some 140 miles off the Californian coast. Formerly the sea-elephants existed there in thousands but were very nearly exterminated till the American influence protected the animals. A bull and two cows were " cut out " of the herd by being fenced off in the manner shown. The fences were then put together as cages, and the animals towed off to a ship.

RESTING AND RAGING EXPRESSIONS OF A TWO-TON BULL SEA-ELEPHANT

When full grown the bull sea-elephant has a well-developed " trunk " (bottom). In repose the expression is calm and the eyes almost tender, but when anything occurs to upset the feelings of this fourteen feet of blubber and muscle the animal leaves one in no doubt as to its state of mind (top). These photographs are of the big bull sea-elephant captured with two cows for the San Diego Zoo, as described in page 740. After a little forcible feeding the animals settled down to about 120 pounds of fish a day between them.

the cachalot, are monsters in themselves, and the writer has seen one taken from the stomach of a whale, that, inclusive of the long tentacles, measured twenty-seven feet.

The sperms roam about in tropical and sub-tropical seas but certain of them also wander towards the poles, and in summer may be seen, for instance, off the west coast of Ireland. These when caught are found to be males and it is supposed that they are young or very old members of a herd that have been driven away by the stronger "bulls."

Contrasted with the dolphins and the cachalot are the whalebone whales, which do not possess teeth but have instead, suspended from the upper jaws, a horny substance that has generally been termed "whalebone" or baleen. This

also belongs the common dolphin itself, with its long beak-like jaws and elegant slender body, and the smaller porpoises of our coasts. Both dolphins and porpoises are not uncommon sights, for they are fond of playing about vessels, but apart from this not much is known about their habits.

THERE are larger types of whale also that possess teeth, but these only occur functionally in the lower jaw. The best known and most powerful whale thus armed is the sperm whale, or cachalot. It grows to fifty or sixty feet in length and is readily distinguished from all other whales by its huge barrel-shaped head. It is a powerful diver and may "sound" for upwards of an hour, during which time it is seeking its prey in the form of squids, animals of the octopus family that apparently exist in tremendous numbers in the intermediate waters of the ocean. Some of these squids, although insignificant in size when compared with

W. S. Berridge

SEA-LION BEHAVIOUR IN CAPTIVITY

The hind limbs of the sea-lion are more adaptable for life ashore than in the other seals. An example of this occurs in the lower photograph in which we have a sea-lion scratching the back of its neck with one of its hind flippers which is as convenient for that purpose as the hind leg of any puppy. When in the water (top) the limbs are used only for steering.

SEA-LIONS ASHORE ON THE ROCKS TO BASK IN THE SUN

Where they are not accustomed to be disturbed sea-lions can be approached if caution be used so that the animals do not become alarmed. In the lower illustration one of these creatures is lying happily on its side so that the sun can shine upon its stomach. Despite the nearness of the photographer the animal has not bothered to move from its comfortable position. Above are some sea-lions basking on the rocks at Santa Catalina Island, off California These are roaring at intruders who have invaded their peace.

material exists in plates that are divided on the inner side into a myriad of bristles. These bristles bridge a space left between the upper and lower jaws when the mouth is closed. In spite of a lack of teeth the whalebone whales are carnivorous in nature and the manner of feeding happens thus : the mouth of the whale, large even for so great an animal, is opened and water taken into it; the cavern is then closed again and the water forced out of it with the assistance of a huge mobile tongue. The bristles of baleen stretching across the jaws thus act as a sieve that allows the sea water to pass through freely but enmeshes the contained animal life. The shrimp-like creatures caught thus are minute in size, and so the whale depends upon the myriads of them that occur in the upper waters of the sea.

To this type belongs the Greenland Whale, which used to be hunted from small rowing boats amongst the icefloes bordering the Arctic Circle, also the Blue Whale, which has been recorded up to a hundred feet in length and may be considered the largest of all living creatures. Like the smaller finner whales which are fairly common off the British coast, the Blue Whale is comparatively active in its movements and so it could not be tackled successfully by the older methods of whaling.

With the introduction of small steam whalers armed with a harpoon gun and working from a shore station, however, such types may now form the bulk of a whale fishery and so the industry is carried on both North and South of the Equator in places as far apart as South Africa, Japan and the Falkland Islands. The oil, or blubber is still the most valuable commodity obtained from these animals, and it is used in quantity for making margarine.

It is the sea, therefore, and not the land that supports the largest of creatures, for an elephant is but a pigmy compared to the greater types of whale, and two main factors that determine this position may be considered. First of all there is the problem of a food supply for such tremendous bulk, and here it has been seen that the whales are most fortunate, for they are able to garner huge quantities of concentrated animal food the like of which does not exist even in the lands of the luxuriant tropics. Thus the sperm whale seeks below the surface of the water for the shoals of succulent squids, while the baleen-carrying whales strain from the sea the countless numbers of animalculae that exist in it. Nor is such "pasturage" easily exhausted, for the whales have practically the whole of the limitless ocean for their journeyings. In this respect they have an advantage over certain gigantic extinct marine reptiles, for these were probably very sensitive to temperature changes and evidently preferred the conditions of the warmer seas.

Then there is the mechanical problem of supporting a huge heavy body. A land animal may be regarded as a bridge with the legs as supporting piers that take the whole weight of the animal and have at the same time to be mobile. In the case of the elephant the limbs are strengthened and buttressed by huge sinews and muscles, but there is a natural limit to this type of structure.

With the whales, however, the support of the body weight does not enter as a distinct problem ; the animal is cradled and buoyed up by the surrounding fluid so that the structure is concentrated upon affording the necessary rigidity for movement through water, a problem that is compatible with great size as may be noticed in the case of shipbuilding. This specialisation of the whale for its particular environment is shown by the manner in which it collapses when on the beach so that the organs are not able to function properly, and in consequence the animal dies.

Apart from the seals and whales there is another group of mammals, the Sirenia or sea-cows, that may be regarded as marine, although its members never venture into the open sea but hug the shoal waters and lagoons of the tropics. They are smaller and more squat than the whales but somewhat similar in appearance, so that in the past they have been confounded with them. The head, however, is rounded, and there is a well-marked neck ; the nostrils are at the end of the muzzle as in most land animals and not on the top of the head, and the skin is covered by a growth of short stiff hairs. Their mode of life and feeding is different also, for they browse upon the luscious marine and fresh-water weeds that abound in shallow water, seizing them in a mobile mouth and grinding them with strong molar teeth.

The two living types of the sea-cow are the manatee and dugong. The latter is found in the Indian Ocean and differs mainly from the manatee in having a forked tail instead of one rounded in shape. The manatee itself is common to the West Coast of Africa and to the east coasts of Central and South America, and is similar to the dugong in habits. It may be submerged for about twelve minutes and then rise for air, breathing quickly several times before descending again.

The nature of the sea-cows leaves them open to attack by man, and it is not surprising therefore that one species, Rhytina or Steller's sea-cow, discovered over a limited area within the Arctic Circle as late as 1741 and slaughtered for its oil, was rapidly exterminated. Thus the question presents itself as to whether the seals and whales also may not succumb to man once a hunt is organized against them either on the grounds that they are competing for fishery "rights," or because they yield commodities of commerical value.

The Greenland Whale has been hunted almost to the point of extermination, and although sperm whales are still harpooned they no longer supply a whole "fishery" but are only taken incidentally. In view of past experience it would seem that some restrictions may be necessary to prevent marine mammals from being over-exploited.

CAPE AND CALIFORNIAN SEA-LIONS, AGILE IN WATER AND ON LAND

Of the Pinnipedes (wing-footed animals) the sea-lions are the most active on land. They can double the hind limbs under the body and use them as legs. They can achieve a kind of shuffling rocking gallop and, not only that, but can climb over rocks and up steep places in the most surprising way. In the water these useful strong hind limbs are not employed for swimming but only for steering. Below is a Cape sea-lion with very large and powerful fore-flippers and above is a Californian sea-lion.

Arthur Brook

PART OF A HERONRY BUILT IN THE TOPS OF HIGH TREES

February is the usual month for the herons to start nest building and this they do in colonies. The site most often chosen is a clump of high trees in some copse or spinney but cliffs and steep places in the sides of hills may be used. The nest, which is built of pieces of stick and is lined with grass, twigs and roots, is broad and flat. Three or four eggs are laid and hatched in about a month and sometimes another clutch is laid two weeks later and hatched among the growing chicks of the first brood.

Chapter LXIX

The Nesting Habits of British Birds

By Edward Step

Author of "By Vocal Woods and Waters"

ALTHOUGH the first birds inherited from their reptilian ancestors the method of propagation by egg-laying, and probably left their eggs to be incubated by the sun, or by covering them with fermenting vegetation, most of them found that a more certain way was to sit upon them, so that the warmth of their bodies might be transmitted to the living germ and the period of development in that way might be shortened, the hatching hastened and the brood protected. The Megapodes of warmer climes still shift, like tortoises, their parental responsibility on to the sun; but whilst a number of British birds retain the primitive habit of dispensing with nest-building, only one neglects incubation, though she—the cuckoo—is careful to foist the task parasitically upon some other bird.

Most of the British birds which have retained the primitive habit of laying their eggs upon the ground, with little or no attempt to construct a nest, are those of the sea-coast, and these chiefly the group that includes the oyster-catcher and the plovers. The first-named handsome bird merely scrapes in the upper part of the shore a saucer-like hollow. In the depression three (occasionally four) buff or olive eggs, all blotched and streaked with black, are laid, with their smaller ends in the middle to economise space. The stone curlew, or thick-knee, lays its similar, but lighter-coloured eggs under much the same conditions, on the upper shore or on a waste running down to it. That the eggs harmonise with their surroundings sufficiently to afford protection is shown by the fact that only two are laid. The sandy-coloured young are quite as difficult to see as the eggs were.

Other plovers, such as the lapwing and the golden plover, make some advance to what we should consider comfort by selecting a slight hollow among inland vegetation and introducing a little dried grass or some dead leaves as bedding. But what looks like the most casual and risky of all egg-laying is that of the guillemot, which has only one large egg to venture, yet places this on the crowded bare stone shelf of a high sea-cliff. The conical shape of this egg, which allows it to roll only in a small circle or to revolve upon its own axis, has been used often as an example of adaptation to situation; and many have believed that it is really perfectly safe upon its narrow shelf.

E. Step

UNPROSECUTED TRESPASSERS

Turtle doves, like all their tribe, build only the simplest of nests and sometimes make them on top of an already occupied nest belonging to another species. Here it is a pair of song thrushes whose home has been used as a foundation.

WE have seen thousands of brooding guillemots on the ledges of a lofty headland alarmed into sudden flight by the horrid screeching of the siren on a pleasure steamer. Most of the eggs spun round without leaving their shelf, but a number rolled into the deep water below. No egg is subject to greater variation than this; and it has been said to be all but impossible to get two guillemots' eggs that are exactly alike. This, however, refers chiefly to colour and markings; but it seems pretty clear that the eggs that rolled into destruction varied somewhat in shape from the normal.

The razorbill, though its solitary egg is smaller than that of the guillemot, less produced at the thinner end and therefore more disposed to roll, breeds on the cliff face, but selects the holes and the broader terraces where there is less need for this structural peculiarity of the egg. Although the bed is stony no effort is made to soften it by a scrap of dry grass or seaweed, though this might be furnished with the greatest ease.

Among our inland birds the nightjar affords an example of egg-laying upon the bare ground, at the most among the natural débris of heather and bracken or the fallen leaves on the edge of a wood. But though laid with seeming carelessness, often quite in the open, the two eggs appear to be perfectly

Peter Webster

GUILLEMOTS ON A NESTING ROCK AMONG THE FARNE ISLANDS

Some birds take much, others very little, trouble over nesting, but of all the casual ones, the guillemot seems to take the prize. This sea-bird not only neglects to make even the suggestion of a nest but deposits its single egg on a ledge in the cliff and holds it, point outwards, between its legs. The egg has the shape of an attenuated pear and will roll only in a small circle. Probably only badly shaped specimens ever fall off the ledge. Here we see one of the colonies in which the guillemots always gather.

secure, for the average wanderer would pass them by as small pebbles, which they resemble closely. So with the two chicks—their covering of sooty grey down renders them invisible among the soft shadows of twilight, when they are left alone.

The grotesque-looking puffin, another of the marine birds that lays but one egg, prefers the grassy summit of the cliff, where in company with a few or many of its kind it either excavates a tunnel of its own or appropriates the burrow made by an adventurous rabbit. Here, though a weatherproof dwelling is provided, there is no pretence of nest-making or the provision of even a slight bed for the white egg or the blackish downy chick.

This utilisation of holes for breeding purposes was, doubtless, one of the earliest devices of the primitive birds, as an improvement upon the open ground laying ; and we see it continued to-day by the splendid kingfisher and the sand martin, though the latter has more refinement and furnishes the tunnel she has made with a bed of dry grass and feathers. The hollows of decayed trees, also, appear to have been seized upon by those early birds that preferred a woodland life. Many of them have kept up the practice to the present day ; and we find the owls inhabiting hollows in more or less ruined tree-trunks, but making no attempt to furnish the apartment with bedding. This is not the only resort of the owls, for they are not above taking advantage of any suitable hole, or even adapting to their purposes the nests of rooks or crows or convenient ledges in barns or church towers and steeples.

Another of these nesters in decayed trees is the wryneck, which has no sense of providing comfortable bedding for its young beyond the scratchable spongy wood already there. The nuthatch, with similar habits, has some building skill, for where the entrance is larger than is necessary it obtains mud and fills up the surplus space, thus probably keeping out some objectionable visitors.

The wryneck's relations, the woodpeckers, which have the same habit of nesting in decaying trees, prefer to make their own entrances and excavations ascertaining by the sounds caused by their tapping bills where the wood is soft enough for their purpose. There a clean, workmanlike hole is cut through the bark and into the interior wood. The only bedding material is such scraps of rotting wood that the carpenter may have failed to clear out.

An advance upon these rude nesting places was made when some birds realized that soft bedding was desirable for such fragile treasures as eggs and for the tender, naked bodies of newly-hatched young. The common tern, nesting in companies on the seashore, lines its depression with a little dry grass or else some fine seaweed.

More definite approach to nest-building is shown by the gulls, which make rude cushions of gathered vegetation, and, the birds being sociable, large numbers of the nests may be crowded together, often touching. One of the most striking and easily seen of these associations is that of the black-headed gull, which breeds inland in marshes and around pools. In their favourite resorts many thousands of the nests may be visible at one view, those on the drier ground mere hollows in tussocks of coarse grass; but in the wetter parts of the area there are beds of flag-leaves, reeds and sedge, lined with moss and other fine material. There the three (sometimes four) heavily spotted eggs—that appear on the table often as plover's eggs—are laid, and later their place is taken by the black-spotted downy chicks that are able to run and swim at once, though they look to their parents for food.

The herring gull and the common gull, which prefer the ledges of cliffs and the tops of high rocks surrounded by sea for their nest communities, make a rough platform of twigs and dry seaweed, lined with grass and seaside herbs, but its flatness and lack of weaving show that it is primitive. In essentials the nests of rook and crow, though built on the higher branches of trees, are of the same elementary type; still, they exhibit a considerable advance, for the foundation of twigs in that of the rook is made firm by the aid of mud, and the cup is lined with grass and soft roots, whilst the crow achieves comfort by the use of hair and wool as well as grass.

JACKDAW and magpie show the same liking for sticks as a strong foundation, with a more comfortable lining of softer stuffs. Though both may be found in trees, the jackdaw has a fondness for ruins and steeples, where it often accumulates an unnecessarily huge pile of sticks before beginning the collection of feathers, moss and wool for bedding purposes. The magpie's root-lined nest is a more finished production, showing greater art in the contrivance of a dome and a doorway, the latter often framed with thorny branches. A vast amount of

A. H. Willford

MALE OYSTER CATCHER TAKING OVER DUTY FROM THE FEMALE

As a rule the oyster catchers lay their eggs in some little cavity among sand-hills or the shingle. There are usually three or four to the clutch, the shells being brownish yellow mingled with dark brown and grey markings. This remarkably skilful photograph was taken just at the rare moment when the father bird had returned to the nesting site in order to take over his spell of incubation while his mate went away to find her food. The female has just stepped off the eggs and the male is just about to settle.

A. H. Willford

NESTS OF ROBIN IN A KETTLE AND OF MISTLE-THRUSH IN AN APPLE-TREE

Robins repay the interest which men usually take in them and the sanctity which they usually enjoy at human hands on account of old traditions by providing a certain amount of amusement. The little birds, besides their entertaining boldness, have a habit of nesting in the most extraordinary places, often choosing some man-made article for a nesting-place. Here is one (bottom) where a robin's nest has been found in an old kettle pitched long ago into a quick-set hedge. Above is a mistle-thrush on its nest in an apple-tree.

Arthur Brook

PROBABLY THE RAREST BIRD IN GREAT BRITAIN: THE KITE AND ITS NEST

It is believed that only about six pairs of kites are now left to breed in Britain and all of these do so in Wales. In view of this it seems surprising that this bird has been recorded as having been common actually in the London streets as late as the sixteenth century, when it was regarded as a useful scavenger. Persecution has dogged the bird, and it was lucky for this specimen that it was a photographer, not an egg collector, who discovered its hiding-place. The nest is usually in a tree and is made of sticks, bones and even rags.

A. H. Wilford

NIGHTJAR ON ITS EGGS, LAID AMONG SOME RUSHES

Two eggs are the common number laid by the nightjar and they are deposited on the ground without any attempt at making a nest. The bird relies on camouflage to make both it and its young ones safe. The nightjar which, despite its unfortunate name of "goat-sucker" lives exclusively on insects, always prefers uncultivated ground for its period of stay in Britain. The nesting site is usually found among bracken or gorse and the eggs are cream-coloured with brown, grey and purple-black markings.

material goes to its construction; and next year the old nest may have additions to fit it for further use. Another crow, the beautiful and suspicious jay, is a far more neat and finished builder. She has no liking for the high, exposed branches of the tree, but builds lower where there is plenty of foliage for concealment. The nest is a deep, thick-walled cup of twigs, lined with soft roots and grass to hold the six or so eggs that are freckled with brown on a pale green ground.

The birds of prey are quite old-fashioned in their nesting habits; the kestrel being content with a hole of some sort, whether it be a hollow in a decaying tree, a rock crevice, the abandoned nest of magpie or crow, or some ruin to which it has access. There it lays its five roundish mottled eggs without any preparation in the way of bedding. The sparrow-hawk improves somewhat on this with a platform of sticks in the higher branches of a tree, and lines it with soft roots. The peregrine makes a similar nest—if any—on high sea-cliffs; but the merlin prefers the moorland, selecting a hollow on a heather-clad slope and lining it with dead heather twigs.

The small group of native doves are users of sticks, but have not got beyond the skill of fashioning a simple platform, the blue rock dove, the ancestor of our domestic pigeon, being content with almost any rubbish with which to fashion a pad on a rock ledge in a sea-cave; and the stock dove sometimes constructs hers under a bush or in a tree-hollow. Each species lays only two white eggs, so they have not much to provide for. The ring dove or wood pigeon and the migrant turtle dove are satisfied with thin open-work, so loose sometimes that the eggs may be seen through it from beneath. That of the ring dove is balanced usually on the bough of a large tree or its sides supported on two neighbouring branches. The turtle dove builds lower down, in tall bushes or young trees. Once we met with it under peculiar conditions. A pair of song thrushes had made their nest between the forking branches of a young tree, and several eggs had been laid in it, when there came along a newly arrived pair of turtle doves eager to begin housekeeping, and these decided that the nest would make a good support for their flimsy floor.

As a rule, the fresh-water birds are not builders of very neat nests, the quantity of material being more obvious than fine workmanship. Most of them take advantage of the cover afforded by the abundant growths of reed, sedge and flag that mark off the firm land from the water; but some resort to holes in the bank, and others come very near to constructing floating rafts, though it will be found that these are based upon water plants growing up from the bottom. The little grebe and the coot make such nests, sometimes on the mud near the margin, but often far out from the shore, when they have the appearance of a free, floating mass of decaying weeds. Closer inspection will show that the upper surface is hollowed and lined with finer material to accommodate the eggs.

Arthur Brook

RAVEN AND CARRION-CROW FEEDING THEIR INSATIABLE YOUNG

Six infant ravens take a lot of feeding and this mass of yelling beaks (bottom) means hard work for the parent birds. Ravens are fond of nesting on sea-cliffs and especially on headlands and begin to nest in February in the south of England. The general mass of twigs is lined with wool and rabbit's fur, and from three to five eggs are usually laid. The carrion-crow (top) is fond of wooded country. The nest is made in April in a tree if possible. Like the raven, the carrion crow likes wool with which to line its nest.

Arthur Brook

TAWNY AND SHORT-EARED OWLS

Hollows in the trunks of dead trees are the favourite places for the tawny owl to lay her eggs, though she does not trouble to construct any nest inside. The lower photograph shows a tawny owl about to enter the nesting hole in an old elm tree. Above is a short-eared owl which has deposited the eggs among some bracken.

The familiar moorhen prefers to nest in the screen of taller aquatic plants, and is not averse from using the lower branches of a tree, the top of a pollard willow or some other unusual situation. Some years ago we photographed a nest on a Surrey pool that, to all appearance, was floating in mid-water. The truth is, a gamekeeper finding one of his pheasant coops in a verminous condition had immersed it in the water for its purification, and on this the moorhen had built her basket-nest of alder twigs, rushes and grass, in which were six chocolate-sprinkled buff eggs.

The mallard, or wild duck, however, prefers to have her nursery at some distance from the water, and sometimes it is very remote, perhaps under a furze or blackberry bush. It is a bulky structure, looking as though put together hurriedly, and composed of reeds or rushes and dry grass; but it is made comfortable by a lining of down plucked from the breast.

But it is among the smaller birds of our woodlands, commons and thick hedgerows that we must look for the most accomplished artificers in this kind; the makers of more or less neatly woven basket-work. These are by no means equal in proficiency, some of the nests looking as though their makers

Arthur Brook

REMARKABLE "CLOSE-UP" OF A JAY UPON ITS NEST

In a deep cup of twigs, thick-walled and lined with grass and roots the shy, beautiful jay, a relative of the crow, lays its half-dozen eggs. The jay keeps to the woods where the trees and bushes grow the thickest and makes its nest either in a bush, or else on the lower branches of a tree where the foliage is thickest. The remarkably handsome head is very noticeable here with its mottled feathers. The jay is busy at nest-making in April but only glimpses of its brown, black, white and pale blue plumage may be seen for this is a very shy bird.

Arthur Brook

WATER BIRDS' NESTS: LITTLE GREBE AND DIVER

What looks like a drift of floating weeds (bottom) is really the raft on which a little grebe has made its nest. The little grebe, or dabchick, lays its eggs between April and August and makes a large nest for so small a bird. But the nest is so well laid among the weeds that only a close inspection reveals it. Above is the nest of a black-throated diver beside a Scottish loch.

were not yet out of their apprenticeship, though following family tradition. They are not all tree or bush-builders, and ground nests call for less elaboration, as we see, for example, in that of the skylark. This bird is as careless as the shore birds and the plovers, nesting on open ground where the roughness of surface is the chief protection. There is better cover in the cornfield, which it selects sometimes, but often the nest is found in the pasture, in a depression made by hoof of horse or ox, and shielded only by rough grass or thistle. It is a loosely twined saucer of grass, lined with similar material of finer texture, and contains about five eggs that are closely speckled with some shade of brown with slight indications of a whitish ground.

The titlark, or meadow pipit, makes a nest of the same type, perhaps deeper, but places it less in the open, preferring taller, closer herbage ; and though dry grass is the chief material, the finer lining includes hair. Several of the buntings select the ground of fields, downs or hedgerows, but the cirl bunting prefers a bush or the low branches of a tree. The nests are substantial cups of grass, fine twigs or roots, with a lining of hair ; and the eggs are white with chocolate streaks that look as though scrawled with a pen, so that one of the group—the yellow bunting or yellow-hammer—is known also as the scribbling lark. The reed bunting differs from the others by selecting marshy situations for its nest, which includes local material as well as grass, and part of the lining may be the soft plumes of the reed.

The nightingale, too, nests on the ground or but little off it. On the edge of the wood, or in the coppice or thick hedgerow, it makes a quite loose structure of which the circumference will consist of

SWAN'S NEST BY THE BANKS OF THE THAMES

Swans build their nests on a large scale, of rushes, reeds and the coarser kinds of herbage. The eggs, which number from three to five in young birds and up to a dozen when maturity is reached, are laid in April or May and take well over a month to hatch. The nest seen here was photographed just below Caversham Bridge, where it stood on an island, or eyot—the favourite nesting site for Thames swans. Nine cygnets were hatched and in the nest were found an old jam pot and a bottle which had contained stout.

Webster R. Carslett A. H. Wilford

YELLOW WAGTAIL, LONG-EARED OWL, WOODCOCK AND CORNCRAKE WITH NESTS ON THE GROUND

April sees the yellow wagtail (bottom left) constructing its nest of grass and moss lined with feathers and hair or roots, in a hollow in the ground, probably in an open meadow, though there it will be wonderfully concealed. The long-eared owl (bottom right) seen above has selected a site among the bracken, and this fine photograph shows the curious barred effect of the sunlight coming between the fern stems on to the bird's back. What looks like a forest when seen from this angle (top left) is really the thick bracken where a woodcock has nested. Woodcocks always nest in the ground in thick cover and line a depression in the soil with a few dead leaves. A corncrake or land-rail (top right) often makes its nest among standing hay or corn.

Arthur Brook

BROODS OF BIRDS OF PREY: SPARROW-HAWK AND MERLIN

Arboreal in general habit, the sparrow-hawk (left) usually builds its nest on the branch of a tree which is close to the trunk. This nest is rather a rough affair of sticks outside and twigs inside, and sometimes the bird prefers to utilise the old nest of a wood pigeon. There are four or five eggs in a clutch and these are laid, as a rule, during the first few days in May. The merlin (right), the smallest of all the British falcons, just scratches a hollow in the surface of the moor where it lives and deposits its eggs—four to six—in there, seldom troubling about any kind of lining. The eggs which are of a deep brown-red or purple-red colour are laid in May.

leaves, usually those of oak, with a lining of soft material in which may be down from the seed heads of plants. The five glossy eggs are tinted dark olive. The nightingale's relation, the friendly robin redbreast, prefers a hedge bank for its not very well compacted nest of dry grass and dead leaves; but though the exterior is loose, there is a comfortable lining of hair and wool to receive the five whitish eggs freckled with red.

The blackbird and the thrush build substantial cup-shaped nests, placed in hedges, bushes and other convenient places that may offer. The builder can be determined at long range, usually, for the black-bird's nest has a more untidy exterior, with often a conspicuous loose length of straw advertising its position. Both, however, are well-woven structures of slender twigs, roots, dry grass, compacted with mud; the blackbird's lined with moss and soft decayed leaves. The thrush's is much neater altogether and, apart from the marked difference in the eggs, can be known at a glance by the smooth plastering of the inner walls with mud and cow-dung.

Arthur Brook

KESTREL NESTING IN OLD HOME OF MAGPIE

The kestrel—which also goes by the picturesque name of the wind-hover—hardly ever makes a nest of its own, but seeks out the vacated premises—empty since last year—of other birds, particularly magpies and wood-pigeons. Otherwise it lays its eggs in holes in the steep sides of cliffs or quarries.

Some of these bush-builders are less efficient as basket-makers; such is the butcher-bird, which gathers a great quantity of material but leaves many loose ends of twigs. The cup, however, is more neatly made and, being lined with soft moss, wool and hair, must be quite comfortable to the five or six chicks that emerge from the whitish eggs that are speckled with brown. Built usually in thorny hedge, among hawthorn, bramble or wild rose, it is neither easy to see nor to get at.

Similar, but neater, cups are those of linnet and greenfinch, securely hidden among the spines of furze or the pungent leaves of juniper; but much prettier ones are elaborated by the chaffinch and the goldfinch, which are more closely woven of finer material, grass and hair combined with moss and wool, decorated outside with lichens; the chaffinch lining hers with hair and feathers, the goldfinch using vegetable down for her smaller nest.

Among these small birds are some that like to have a roof over their heads when sitting and for the greater comfort of their young; so the walls of the cup are extended upwards until they form a dome over the entrance. Such structures are found on the nests of willow-warbler, wood-warbler and chiffchaff. An advance upon this style is seen in the nests of the wren and the long-tailed tit; the former a large affair for so small a bird, built of moss, grass, lichen, lined with feathers and the entrance near the top. The long-tailed tit uses kindred materials, and the nest is so decorated with tree-bark lichens that it harmonises with the branches among which it is placed. Moss, wool and spiders' webs go to the construction of the body, and the interior is made luxurious by a profuse lining of soft feathers for the numerous red-speckled tiny eggs. Neatly woven, too, is the very deep cup of the reed warbler, though it is without a roof, and the material the long leaves of reed and grass, so contrived that several strong upright stems of the reed pass through its walls, supporting it well above the water or marshy ground; softer material such as hair and feathers are used for lining.

Chapter LXX

Animal Marksmen Trained by Nature

By Frances Pitt

Author of "Waterside Creatures"

SOME wonderful marksmanship is found in Nature, many creatures attaining an uncanny accuracy of aim, so that they seldom miss the quarry towards which they direct their efforts.

I was thinking as I penned that line of birds that convert themselves into living bullets, the peregrine falcon that hurtles from the sky upon its prey, or the gannet which dives headlong into the sea after a fish. They are indeed marksmen, Nature having endowed them with wonderful skill. Only a short while ago I was upon an island off the Welsh coast, watching the bird life of cliff and sea, scarlet-beaked choughs upon the rocks, smart oyster-catchers tripping over the shingle, and patrolling gulls flying over the waves. Then a big white bird came into view, its large size and black-tipped primaries proclaiming it that fine sea bird the gannet, or solan goose.

To and fro it flew, up and down the sound that separated the island from the mainland, evidently scanning the green water beneath and on the look out for a fish. It was flying at a fair height, quite thirty feet above the sea, when it closed its wings and dropped headlong, diving straight into the waves and raising a fountain of spray. It was a most spectacular plunge. The force of its dive must have carried it far under water, yet such was the marksmanship of that gannet that when it presently bobbed up to the surface it carried, as was to be expected, a fish in its beak!

There are some other sea birds that fish in a similar manner. I mean the exquisite little terns, or sea swallows. These seemingly fragile, sprite-like birds, that look as if too delicate and dainty to withstand the buffetings of life, let alone the rough waves, float about on their long white wings watching the sea beneath them, and the moment they catch sight of a little fish they plunge in after it.

I was watching some lesser terns one day (the smallest and most delicate looking of the tern tribe, mere wisps of birds that might be blown away by a rough breath) that were flying to and fro where a north country river joined the sea. The sun shone on the little white fairy-like birds, on the quiet river, and the golden sandhills in the background. As I watched them one of the terns dropped, falling as a bolt from the blue sky straight into the river—it had seen a fish! Again an accurate aim was rewarded, the tern emerging with a small fish. But as regards a living bullet what bird or beast can surpass the falcon? Marvellous indeed is the marksmanship of the peregrine, as was well appreciated in days gone by, when hawking was an everyday pursuit, and people used a falcon as they now use a shot gun. Yet before speaking of the use to which a hawk's skill can be turned, let us consider how the bird exhibits its accuracy of aim under natural conditions.

W. S. Berridge

GETTING READY TO SPIT.

To spit with accuracy is a feat for which the llama inherits a special ability. Incautious visitors to Zoological Gardens have suffered unpleasantly from the results of the llama's displeasure. Here we see one getting ready to "fire."

A PEREGRINE makes use of its powers of marksmanship in the following way. It gets up a good height in the air, attains its "pitch" (to use the language of falconry) and "waits on" until it sees a pigeon or some other bird beneath it, when, flying hard in the direction of the quarry, it attains great speed that culminates in a lightning dive, when with closed wings it hurtles down upon the bird, in that meteor "stoop" which the eye can barely follow. Slight indeed are the chances of the quarry, which has no time to turn and dodge. When the quarry has been knocked senseless and has fallen earthwards in less time than it takes to tell, the marksman, turning about, shoots down and upon it, to seize the victim as it lies on the ground.

A falcon never or hardly ever grabs its quarry in the air. As it stoops upon the shifting target, it hits the bird with its clenched foot, of which the great talon of the hind toe makes a steel-like keel, and so great is this blow that the victim is generally killed in mid-air, falling in a cloud of feathers.

Dr J. S. Crainer

MONKEY MARKSMAN OF THE TREE-TOPS

Being possessed of hands the monkeys have learned to throw, a trick no other animal possesses. The monkey's habit of throwing coconuts at enemies on the ground has been turned to man's use. This monkey is trained to pick the nuts for his master.

In the days of old, as I have remarked above, this wonderful combination of wing power, accurate eye, and foot work, was brought into human service, hawks being tamed and trained and made use of, these bird marksmen exercising their incomparable skill for their masters' amusement and profit.

At this point the reader will probably enquire how far this skill is instinctive and untaught, and how far it is acquired through experience. There can be no doubt that every young hawk has an inherited aptitude to hurl itself at birds, for I have taken eyasses from the nest, hand reared them so that they were deprived of the old hawks' example, and watched their behaviour the first time a small bird flew up near them, when they dashed after it in headlong pursuit. ("Eyass" is the term that was used in olden days to denote a nestling hawk.) But though the young hawk has the desire to chase imbedded in its mental make-up, it has much to learn, and, as will be remarked later in this article, the pursuit of an unskilled eyass is a different matter for the quarry to the determined drive of a cunning old hawk, with her experience of hundreds of kills behind her.

Now to turn to other skilled marksmen, whose art consists in the use of a spear, we find some fine performers among the fisher folk, to wit the heron and the kingfisher. These two most dissimilar birds have one thing in common, namely, that they make a living by spearing the fish of our rivers and brooks, but how differently they do it, though both exercise art of the highest degree.

Let us take the heron first, that lone grey bird which flaps so solemnly across the sky, to plane gently down and alight by the woodland stream, and stand tall and straight on his stilt-like legs regarding the babbling brook. Then with gentle gliding steps he walks into the water, picking his way between the green liverwort grown rocks, to where the stream deepens into a placid pool. Still he moves forward, a cautious step at a time, keeping to the shallow

W. S. Berridge

POISON SPITTING COBRA FROM AFRICA

Snakes have unpleasant ways of disabling human beings, but the ringhals, or spitting cobra, from Africa is the worst. With dreadful accuracy it can project its poison into one's eye from a distance of some feet. The photograph shows it ready to spit.

James

CHAMELEON WAITING THE CHANCE TO SECURE ITS PREY

No swordsman ever achieved so rapid a thrust with the rapier as the chameleon achieves with its long and lightning tongue. Here (bottom) we see the position taken up when the animal is about to secure an incautious insect. The tail is coiled securely round a branch and the tongue is just beginning to come out from between the jaws. The upper photograph shows a group of these creatures upon a branch, meditating over their last meal and the possibilities of the next. Notice the different postures.

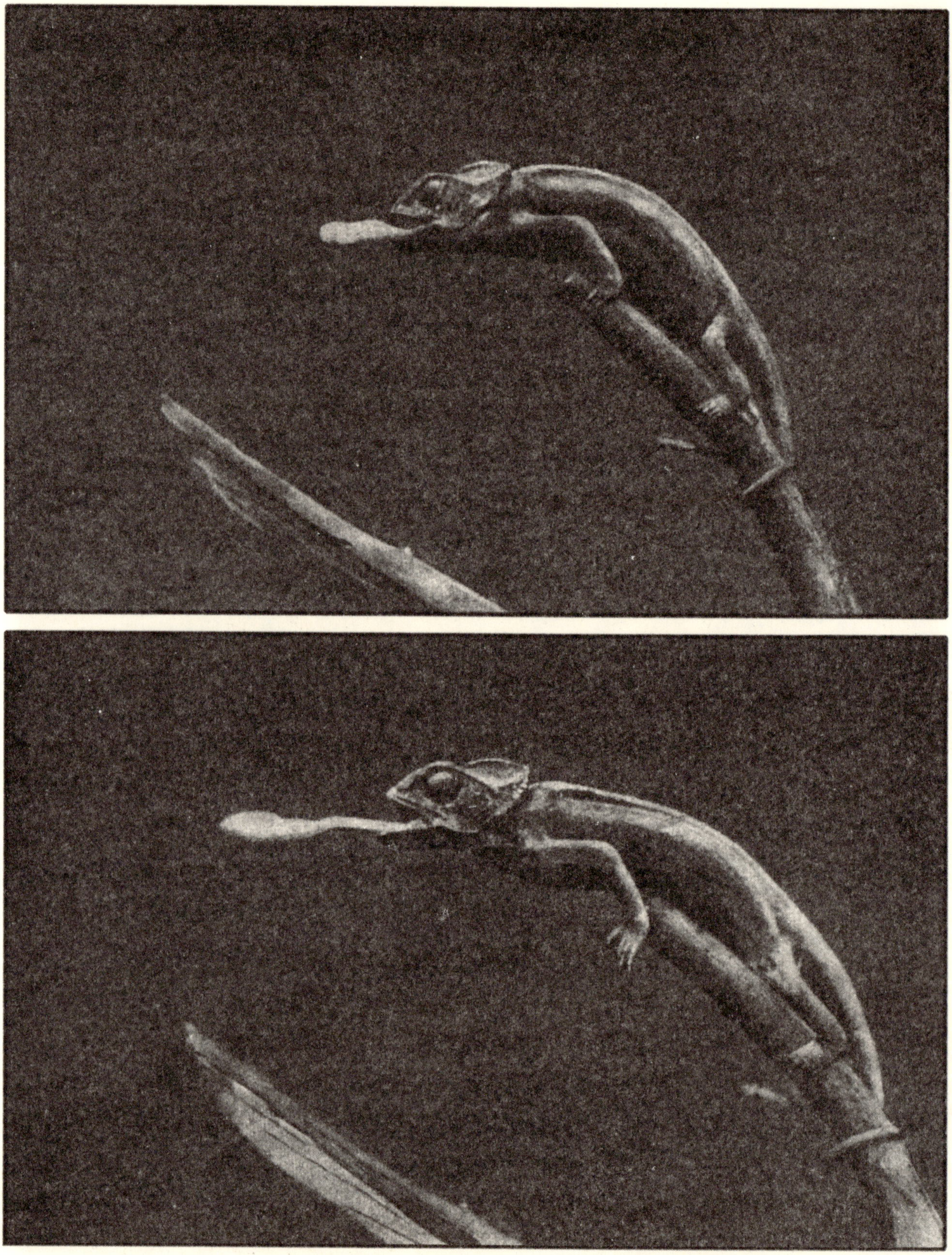

CHAMELEON PREPARING TO SHOOT ITS TONGUE AT AN INSECT

When not in use the chameleon's most remarkable tongue is folded away within a special sheath which is located in the dilatable skin of the chin. The little reptile remains immobile for hours on a branch until some fly or other insect settles within range. Then the chameleon, rolling one of its eyes—which work independently of each other—until it has focussed on its intended meal, protrudes the fat tip of its tongue from between its jaws, takes aim and then, with a movement too quick for the eye to follow, the tongue flies out and the insect is struggling on the end. The end of the tongue is covered with a sticky substance.

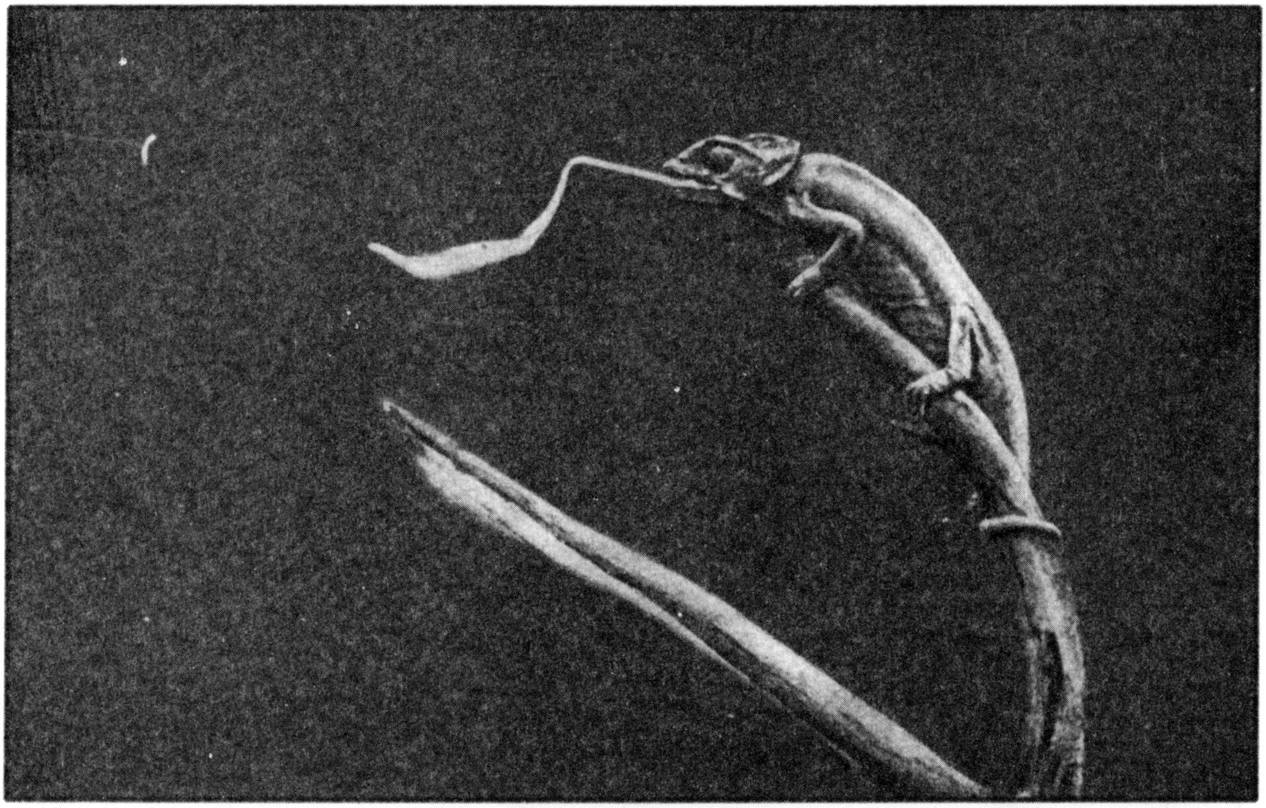

REMARKABLE TONGUE OF THE CHAMELON IN OPERATION

Only the camera can record the different stages in the lightning movement of the chameleon's tongue when it "shoots" at an insect. Compared with the size of the little reptile the tongue is of an enormous length and can capture an insect six inches away. The true home of the chameleons is Africa and the adjacent islands, though there are some species found in Spain, Southern Arabia and India. They are not difficult to tame if allowed to become used to new quarters in peace, and will live in a greenhouse, where the wonderful action of the tongue can be studied at leisure as long as there is a supply of flies available.

ARCHER FISH AND ITS INSECT TARGET

This drawing depicts the method by which the archer fish obtains an insect that would otherwise be out of reach. It takes water into its mouth and spits a large drop into the air, aiming so accurately that the drop strikes the insect and knocks it into the stream, when the fish can easily snap it up.

water and scanning the current with keen yellow eye. Hardly moving, looking more like a grey stick than a living bird, he stares and stares—a shadow in the water, the heron's beak flashing quicker than the eye can follow, and the fisherman has speared a trout. It gleams for a moment in his yellow beak, its spotted sides distinct, then by some sleight of hand (or rather of bill!) he turns it about, so that he holds it head first, gives a gulp, and all that remains of the fish is a lump sliding down the heron's grey neck. He remains quiet and thoughtful for a moment, then gives himself a shake, looks about and wades forward ready to spear another trout.

And what of the heron's competitor, that other expert with the spear, namely, the kingfisher? Well, his methods are different, but equally efficient. Here he comes, darting down the watercourse, flashing by like an electric vision on the wing, a dazzling flash of blue-green light that makes the observer blink, and marvel that such gorgeous colouring should appear in the sober landscape of a European river.

Such hues are tropic in their intensity. But for all his beautiful plumage the kingfisher is no amateur at his trade. He turns aside, darts up to a branch that overhangs the river, and vanishes. He is there all right in reality; he has but turned his chestnut breast to the river and the observer, and his iridescent blue-green back to the overhanging trees. Thus, the kingfisher sits, perfectly still, scanning the flowing water passing beneath, and watching for minnows and little trout. Suddenly he dives headlong in the stream, raising a shower of spray, from which he rises, a small fish in his beak, to fly back to his perch. He beats the fish on the branch, thumping it until it wriggles no more, when down it goes, head first.

So far we have been discussing birds that are skilled marksmen in capturing their quarry, but what of creatures that aim in aggression or self-defence? Under the former heading we must place the llama, and the latter the fulmar petrel. Both have the unpleasant accomplishment of being able to spit straight at a mark. For this reason it is well to beware when visiting the llama house at the Zoo, as these animals can and will spit at people who do not please them. Whether this ability was useful to the llama and its wild relatives in its South American home is a question I cannot answer, but undoubtedly the fulmar petrel benefits by its mastery of the art of spitting. The fulmar petrel, a graceful sea bird, white, with greyish mantle and wings, was once confined to the lonely island of St. Kilda out in the Atlantic, but has spread and increased, so that it now occurs on the great cliffs of the Orkneys and Shetlands, and even upon the Scotch and English coasts. Very lovely it is to watch, so graceful and dove-like upon the wing, so soft and motherly as it broods its solitary chick on some ledge of a tall cliff with the waves thundering far below. Of all gentle creatures it looks the most inoffensive, and so it is—as long as it is not disturbed, yet the person who incautiously approaches the brooding bird is likely to receive the oily fishy contents of its crop full in the face!

W. S. Berridge

SKILFUL ARCHER FISH OF EAST INDIAN AND AUSTRALASIAN WATERS

From the East Indies, Northern Australia and New Zealand comes the archer fish which haunts estuaries on the look-out for flies. A number of fishes have resorted to the habit of leaping bodily out of the water to get their food, but the archer fish has improved on this method by turning its mouth into a kind of blow-pipe as illustrated in the opposite page. This is a very thick-set fish with the spiked dorsal fin placed far back. It can spit with unerring aim to a distance of about five feet.

To return to animals that use their marksmanship to gain a living, the common, familiar frog is not a creature that would be expected to have a "straight eye" or be good at hitting a mark, yet it can aim well, and shoots with its tongue. It is fond of insects and small grubs, coming out to look for them in the twilight It stalks them carefully, getting as near as it can, then out shoots its tongue, hits the fly or grub, and being sticky at the tip retains the victim, which is drawn back to be munched up, when its captor goes on and looks for another.

The chameleon practises the same art, but it does not go and look for things, instead it waits for flies to come to it, sitting immovable for hours until some unwary insect wanders within range, when out goes the tongue and in comes the fly. The lightning flash of that elastic, but deadly accurate, tongue issuing so suddenly from the seemingly lifeless little reptile is one of the many amazing things in Nature.

At this point it will probably be asked whether any animals use weapons with which to hit a mark, or whether the use of weapons is purely human? Monkeys may at once be cited as using stones, sticks and nuts, to throw at foes, picking them up and hurling them with a considerable degree of precision. In Professor Kohler's "Mentality of Apes" we find a most entertaining description of chimpanzees employing sticks as weapons, but chiefly in play and mock warfare, when they used a staff quite effectively. The writer, however, makes no mention of his monkeys throwing their sticks at anything, but only refers to the weapons being brandished and used for hitting objects, including the other monkeys. Yet some of the smaller monkeys throw sticks and nuts quite freely and with an accuracy most unpleasant for the target. But then their hands are well adapted for grasping and throwing, as indeed are those of all the higher apes.

In considering the marksmanship of mammals other than the apes, it must be remembered that the use of weapons is debarred them. They have no means of using them. It takes a hand to grasp a stick or a nut. In considering human achievement, how much do we owe to our efficient hands? Perhaps the human brain would have overcome the handicap of less efficient members, but it is certainly the hand, with its thumb and fingers, to which we owe our mastery of the world, to say nothing of the art of throwing things and aiming accurately at a mark. Hence it is that we have to turn to the monkeys, with hands somewhat of the pattern of our own, to find even limited ability in throwing things. Most other creatures have to turn themselves into a missile. The falcon, already spoken of, is a good example of a living bullet, when it hurtles down from the heavens on a pigeon or other bird. Crocodiles aim with their tails, using this appendage to sweep the

E. N. A.

WHIP-RAYS SPEARED ON A BEACH OF NORTH-WEST AUSTRALIA

From time to time there have appeared before the public certain men who were such experts at using a stock-whip that they could flick the ash from a cigar without touching the smoker between whose lips the cigar was held. The whip-ray has developed a tail much like a stock-whip save that the end is armed, in some cases with a poisoned sting, and in others with a sharp spine, either weapon being very deadly in execution. Here are a number of these unpleasant creatures which have been speared by some natives of north-western Australia.

victim off its feet; horses strike with their feet, kicking in aggression or self-defence; the giraffe, too, can give a shrewd, well-aimed kick at anything attacking it; and then there are the members of the cat tribe, the household puss, lion, tiger, and so on, whose marksmanship consists of accurately springing or hurling themselves upon their quarry.

Apropos of the tail being used to hit things, perhaps the most remarkable development of this special branch of marksmanship is that found among the sting-rays. The sting-rays are fish allied to the sharks and skates, but of weird aspect, most species having a flat, bat-like body with a long, thin tail, the latter being like a whiplash. This whiplash is, however, a most efficient weapon, being armed in some cases with a poison spine, that deals death to the victim, and in others with a serrated dagger that is equally deadly and efficient.

But writing of marksmanship among the fish, the first prize-winner for real skill, even if that skill be an inherited aptitude rather than acquired art, is undoubtedly the archer fish. This fish, found in Java and Sumatra, has the extraordinary habit of spitting at insects sitting on the leaves and twigs of trees overhanging the water, and thus knocking them into the river, where they are easily accounted for. It does not aim directly at its quarry, but over it, so that the drop falls upon the insect from above, and knocks it off its resting-place.

To jump from fishes back to the higher mammals there is the elephant, which is no mean exponent of the art of "shooting straight," that is, when there is a supply of water at hand and it can fill its trunk and blow the contents with some force in the desired direction.

Once, at the London Zoo, I saw two men teasing an elephant in the Elephant House by holding a bun towards it and snatching the offering back just as the animal was going to take it. This went on for some minutes, until the elephant, evidently realizing it was being fooled, swung on its heels, and went to the back of its stall. Here there was a tap, from which water fell in slow drops. The elephant placed the tip of its trunk beneath the tap and stood for some time collecting the dripping water. Then it turned about.

The young men were still there watching the elephant, but had evidently no expectation of what was about to happen, though an instant later they were able to appreciate how accurately an elephant can aim. Up went its trunk, and out shot the water. Straight it came into the faces of the culprits, running down their necks, and all over them. They decamped in a hurry amid the jeers of the onlookers, while the elephant watched their inglorious departure with a satisfied air. Certainly it had managed to hit its target all right!

CHAMELEON, A FINE MARKSMAN, "SHOOTING" ITS PREY

With its tail curled securely round the branch on which it stands, the chameleon takes careful aim at a fly which has just settled on a leaf some inches above the little reptile's head. Then comes a lightning movement, too quick for anything but a camera lens to follow, the long tongue shoots out, hits the fly unerringly and, by means of its sticky end, secures the chameleon another meal.

W. S. Berridge

THRESHER OR FOX-SHARK WITH A HIGHLY SPECIALISED TAIL

Threshers, also called fox-sharks, feed on fish that swim in shoals, such as herring and mackerel, and obtain their prey by swimming round a shoal, threshing the water with the tremendously developed upper lobe of their tail fins. The shoal is in this way induced to take up a more compact formation when the thresher can the more easily dispose of the frightened fish. Threshers are also said to attack whales by lashing them with these long tails. Below is a photograph to show the action of the tail, and above is a specimen caught in a net.

E. J. King

PEREGRINE FALCON THAT TURNS ITSELF INTO A LIVING BULLET

In its huntingt he peregrine flies upwards until it has attained what is called its " pitch " and then " waits on " till it sees some suitable prey flying below. It sets off in pursuit and, judging the distance and its own and its quarry's speed to a nicety, shuts its wings and falls like a stone. The impact when the peregrine strikes its mark is calculated to kill even a large bird, the fatal blow being really a kick with the clenched claw, which is armed with a murderous talon on the hind toe.

Elephants are exceedingly skilful in shooting thus at a mark. But whether their skill could be greatly enhanced by suitable training is another matter. Probably it could, because however complete the instinctive outfit of any creature, it is obvious that practice helps it to co-ordinate eye and muscle and improve its performance. This is why play is of such importance in the life history of birds and beasts, the playful repetition of actions smoothing out hesitations and imperfect correlations, so that when the need arises the creature can behave appropriately with the perfect response of much practice, to say nothing of confidence in itself.

I have described at some length that living bullet the falcon, which hurls itself from the sky with unerring aim at its quarry below, but as was pointed out at the beginning of this article, the young and inexperienced hawk has to acquire that skill. It has a feeling, a " notion " for the work, otherwise an instinct to stoop at birds, but it only attains the perfect skill of its parents after many days of play with its brothers and sisters and of unsuccessful attempts to capture birds.

WHAT a fine sight is afforded by a party of peregrines, the young birds swooping, turning, wheeling and diving, driving at one another from aery heights, dashing up again into the blue, and practising, in the joyous light-hearted abandon of the young, every trick of the wing that will lend them mastery of the air and endow them with the strength, to say nothing of the certainty of eye, to dart from the clouds with the unexpectedness and velocity of a meteor and strike a mallard or grouse headlong into the heather.

Thus we see that the ability to hit an object, in one word, " marksmanship," is of supreme importance to a number of wild creatures, and that the skill of the finished master is in many instances only attained after a more or less lengthy apprenticeship.

Chapter LXXI

The World of Make-Believe

By W. P. Pycraft

Author of "Camouflage in Nature"

ALTHOUGH a vast mass of material has been garnered bearing on the theme of the coloration of animals, there is, as yet, an immense amount of work which must be done by observers in the field before we can penetrate further into some of the more puzzling aspects of coloration where this is associated with behaviour which seems to imply a conscious exploitation of bodily peculiarities of colour and structure. And this, too, where we should least expect to find it—among lowly creatures like insects, crustaceans and molluscs.

In surveying cases of this kind we have also to consider, as might be expected, a number of "borderline" cases, and these should be very carefully examined for they may afford a clue to the more complex instances presently to be reviewed.

In Chapter III of this work mention was made of caterpillars which have acquired an extraordinary likeness to twigs of the plant on which they live; but this likeness holds only so long as they behave like twigs; that is to say, so long as they stand out stiffly and absolutely without movement, an attitude which they must maintain throughout the whole of a long day. We have no reason to suppose that they are afraid to move, but rather may safely conclude that they have become adjusted to feeding by night. Nevertheless, we cannot help asking what agencies have been at work to change so materially form, coloration and habit, and to blend the results together to produce this startling likeness between a living body and a "dead" twig.

No less remarkable is the case of caterpillars of the small moths of the family Psychidae. They are known as basket-worms, and this because they live in curiously wrought tubes recalling those of some species of caddis worms. In one species this case is made of short rods of wood, arranged in the form of a spiral; another, Helichopsyche, spins a case of silk so like that of a snail-shell in form and texture that specimens have actually been sent to the British Museum as snail-shells! We cannot for one moment suppose that one of the caddis-like caterpillars had deliberately imitated the case of the caddis worm, for it would be useless to imitate the semblance of an aquatic creature climbing about on land: nor is it any more likely that the other imitated the shell of the snail. These two cases of resemblance are purely accidental; but they are none the less puzzling.

J. J. Ward

BUTTERFLY MAKE-BELIEVE

While its enemies are busily searching for their food the green-veined white butterfly is pretending, with wonderful success, to be part of the blade of ribbon grass on which it has settled. How closely it harmonises with its background may be seen.

THE art of make-believe, again, is practised by a large number of very different types of animals which in face of peril feign death. And in some the whole form of the body seems almost to have been shaped to lend colour to this piece of deception. The stick-insects, for example, even while they are at rest are hardly to be distinguished from among the surrounding twigs. Remove one from its hiding place and hold it in the hand, or lay it on the ground. Throwing the legs straight forwards to continue the line of the long, slender body, it remains absolutely motionless for a minute or two. Then, when it seems to imagine the enemy has gone, it slowly gets up and walks away! This is a well-known trick of the opossum—hence the phrase "playing 'possum."

The land-rail, again and again, has deceived sportsmen in this way. Taken but slightly wounded and brought by the dog to its master's feet, it may be held in the hand apparently lifeless, with head, neck, and legs dangling limply but if presently put down on the ground it will slowly raise its head as if to see whether the coast is clear, then instantly dart off. Its cousin the water-rail is no less

H. Bastin

skilful in playing the "artful dodger." When we find this ruse practised in highly organized vertebrates we interpret the behaviour as due to a fairly well developed "emotional complex." But it is by no means so easy to understand precisely similar conduct on the part of insects whose brain is known to be of an exceedingly simple type.

Birds will also on occasion feign wounded, as when they desire to draw away men or other formidable enemies from the vicinity of their young. By affecting a broken wing the bird induces the enemy to give chase, and with the prospect of an immediate capture. But as soon as a sufficient distance is judged to have been placed between the "wounded" one and her young she flies off, and leaves the young to save themselves by scattering and trusting to their protective coloration.

BEHAVIOUR of this kind we never meet with in the more lowly types such as the insects, for they rarely take any active part in the care of their young. But we cannot avoid taking these cases of the behaviour of birds and beasts into account when we are considering a like behaviour on the part of, say, the stick-insects. The butterflies and moths seem to show a peculiar aptness to assume a likeness to foliage. The wonderful case of the butterfly Kallima has already been cited in Chapter Three. But surely the case of one of the British butterflies is no less striking. This is the Duke of Burgundy fritillary, wherein the caterpillars, just before passing into the chrysalis stage, group themselves together in little clusters of three or four at intervals along a stem of the food-plant, so that when they have finally assumed the chrysalis condition they take the form of a series of leaf-buds.

Yet this astonishing behaviour is surpassed by one of the hemiptera, *Flata nigrocincta,* a species which has no name in common speech. The

J. J. Ward

INSECTS IMITATING LEAVES

While some creatures make believe to be something else to save themselves from being eaten this group, representing four of a species of mantis from tropical America (bottom), imitates a bank of leaves in order to be able to eat. Some unsuspecting victim will pay with its life for its mistake. Above is a Ceylon green leaf insect.

record of its strange story we owe to Professor Gregory, who, during his exploration of East Africa, found himself one evening in the woods by the Kibwezi river when his attention was attracted by a large, brightly coloured flower—like a foxglove, and below the inflorescence he noticed some small, white, fluffy patches, apparently of a lichen which does not usually grow on flower stems. To examine these further he tried to pull the flower-head towards him with his stick, when, to his amazement, the "flowers and buds jumped off in all directions"!

He waited till his friend just behind him came up, when, pointing to some similar clusters close by, he quietly asked him if he knew to what genus they belonged. The friend replied that though he had seen them before he had never examined them, and at once put out his hand to pluck a specimen. The moment he touched it the same uncanny destruction of the "flower-head" took place.

In the arrangement of this "flower-spike," the green bud-like forms were at the top of the stem and the pink flower-like insects below, while the tufts of apparent lichen turned out to be the larval stage of the same insect, but of a younger generation. The members comprising the "flower-head" then are "dimorphic," that is to say, they present two forms, one green and one gaily coloured. But we are still at a loss to know what determines their resting positions, for we can scarcely suppose that they deliberately sort themselves according to their coloration before settling down. Professor Gregory, in the hope of discovering an answer to this riddle, waited for some time to see whether they would return and take up their original position, but in this they disappointed him.

H. Bastin

FLYING INSECT PRETENDS TO BE A ROTTING TWIG

With wings spread the insect dnomotis is palpably a flying creature going about its business, but when it settles (top) it ceases, apparently, to be a living thing at all and becomes a piece of decaying twig. As it chooses its alighting places to suit its own peculiarity, dnomotis is hard to find when at rest. It is difficult to believe that a flying thing could look so dead.

Yet one more case of insect behaviour which seems to show a conscious effort to deceive. This is furnished by a black and white West African moth, *Deilemera actinorii*. When spinning its cocoon it covers the outside with little whitish

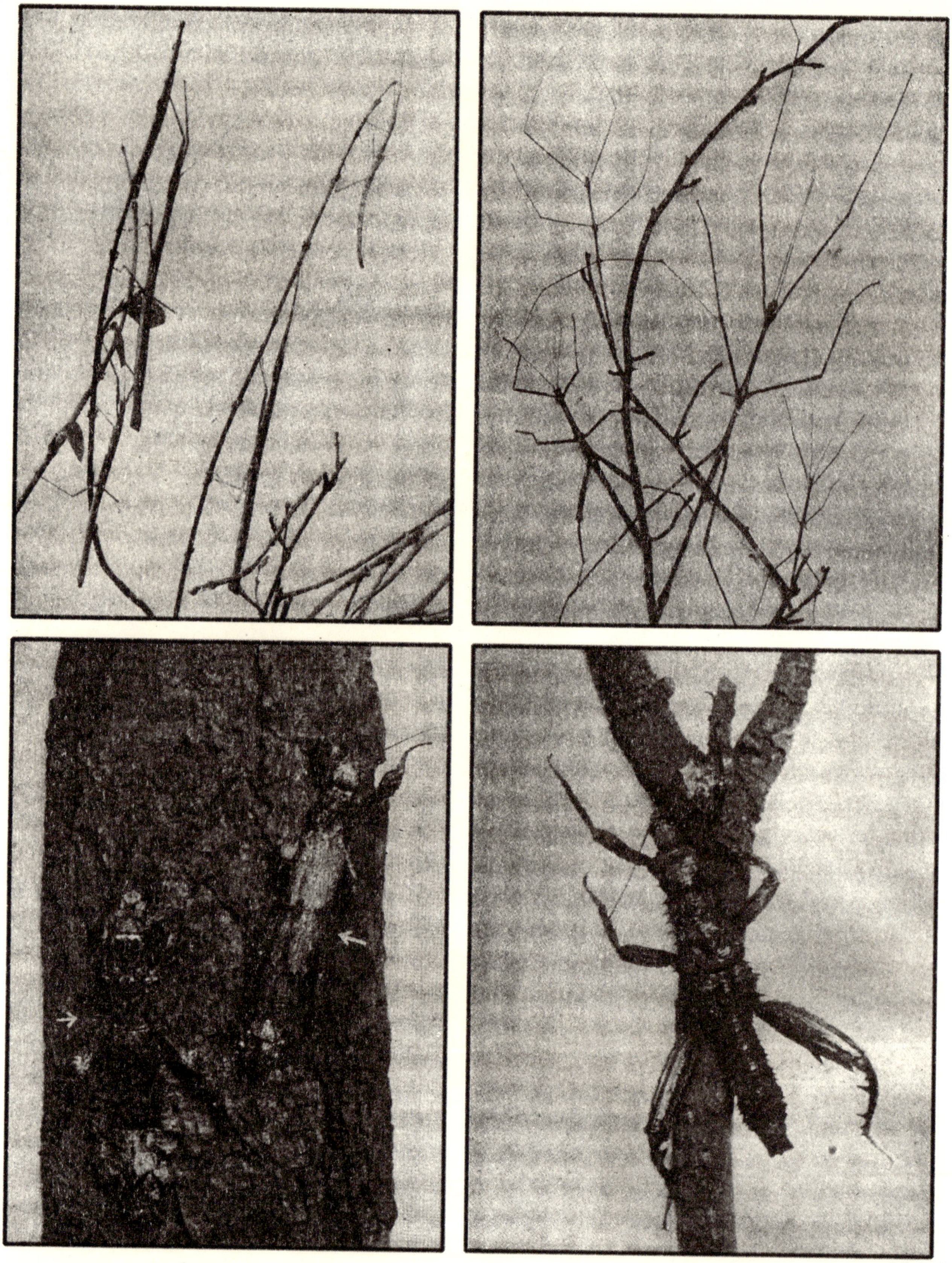

J. J. Ward

STICK INSECTS OF AUSTRALASIA, INDIA AND THE PACIFIC ISLANDS

All the stick insects do not live among the slenderest branches of the trees. Some live upon the bark round the main stems like the Australian specimen seen above (bottom left). When this creature is clinging to bark with its wings folded it is well-nigh invisible as an insect. Of the stick insects that live up to their name more than the first example, we have here (bottom right) a New Zealand prickly stick upon a dead branch, (top left) five Indian stick insects among twigs and (top right) a quartette of Pacific Island insects.

J. J. Ward

CREATURES THAT CUNNINGLY SEEM WHAT THEY ARE NOT

Some of the deceits practised by caterpillars are astonishing. Those of the lobster moth (bottom left), for instance, while resting, make their full-fed bodies resemble the dry parts of the leaves on which they have been feeding. Larvae of the swallow-tailed moth build up an amazing likeness to twigs. Two of the six seen in the bottom right-hand photograph have joined themselves together so as to be like a forked twig. Above are (top left) a Brussels lace moth caterpillar resembling lichen and (right) comma butterflies like withered leaves.

H. Bastin

Martin Duncan

VEGETABLE MIMICRY IN GRASSHOPPERS

Grasshoppers put their trust in something more subtle than the strength of their legs, great as is the power of their limbs for making huge jumps. The lower photograph shows a grasshopper from Borneo which survives by virtue of its resemblance to a leaf while above is a short-horned grasshopper on a grass blade whose surface is copied by the insect's skin.

balls simulating the cocoons of parasites which feed upon the tissues of the helpless body within. Birds discovering such apparently infected cocoons pass them by without more ado, believing that nothing remains but empty husks. Thus, by simulating a parasitised cocoon they secure immunity from attack!

Here we seem to have evidence of foresight and cunning of a very high order. But let us be careful about jumping to conclusions. The cocoon, it is to be remembered, is spun by a caterpillar. It is a task performed without previous experience, and there can certainly be no means by which the spinner can see the result of its labours. Nor can it know that presently it is to emerge from that cocoon as a moth—provided it can succeed in bamboozling the birds! In Chapter Three I described the extraordinary behaviour of the spider crab, which laboriously hooks pieces of seaweed on to its shell until it matches the weeds by which it is surrounded. Its strangely purposeful actions seem to demand that we should regard them as conscious efforts to deceive its enemies. This conclusion is difficult to adopt, and it becomes still more so when we recall that the caterpillar of the blotched emerald moth directly it emerges from the egg proceeds to cover itself with pieces of its food-plant, attaching to each a film of silk, and fastening them thereby to hooked spines with which the body is studded. On emerging from its hibernation it uses oak-leaves.

Now, it is hard to regard this apparently deliberate piece of camouflage as the outcome of conscious effort, for what can a newly emerged caterpillar know of its surroundings, and more especially of the ruthlessness of its enemies which must promptly be hoodwinked? Its near relation, the Essex emerald, behaves in exactly the same way. But curiously enough other

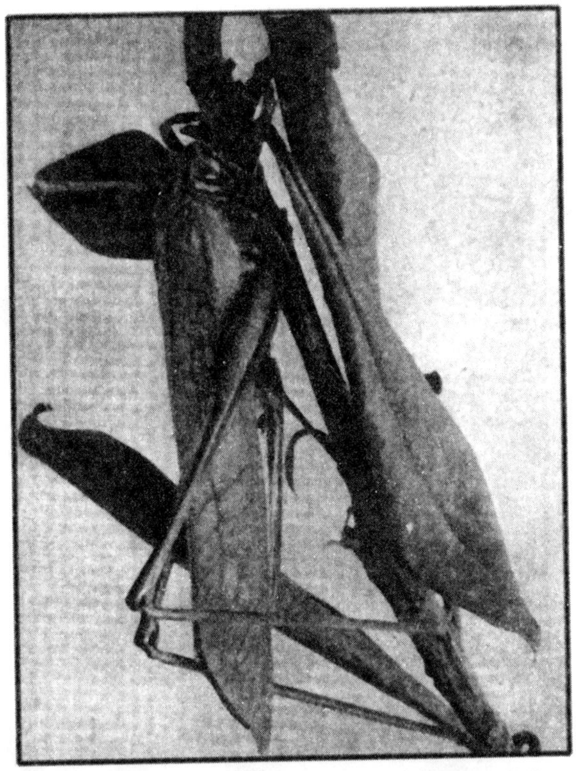

H. Bastin

EXTRAORDINARY DIFFERENCE OF APPEARANCE WHEN RESTING AND FLYING

The right-hand photograph is of a long-horned grasshopper clinging to a leaf, while the left-hand photograph shows the same insect in flight. The two illustrations are hardly recognizable as being both of the same insect, so extraordinary is the power of mimicry while the creature is resting. The grasshopper's very life depends upon this mimicry. While at rest upon a leaf it is much better that it should devote itself to rest secure in its wonderful pretence of being a part of its resting-place than use up energy upon constant watchfulness.

species of emerald moths practise no such subterfuge, but trust to their protective coloration and their habit of posing as pieces of twig.

The larva of a small Hemirobiid insect—one of the Neuroptera, having no name in common speech—adopts a most extraordinary and gruesome device to protect itself. This consists in attaching the empty bodies of aphides which it has sucked dry to the long hairs of its body, forming with this material a sort of shell which looks merely like withered vegetable matter till it is closely examined.

Among the caddis-flies we find the larvae constructing most cunningly fashioned tubes for the protection of their soft bodies. Each species builds after a different method. One spins a shell, coiled like a snail shell, very like that of *Helicopsyche,* already described. Another builds tubes of coarse sand-grains, another of quadrangular stones. Short lengths of weed, bound together, and looking like little bundles of sticks attached one behind the other, or short pieces of coarse weed irregular in shape, are also used. Some use snail shells, and are quite indifferent as to whether the shell is empty or not! Finally, some use a whole leaf which has fallen from some bush overhanging the water. So that while the wearer is resting it looks just like a newly-fallen leaf lying at the bottom of the water!

It is curious to note that the sauba ant of Brazil, when it takes its walks abroad, holds a leaf, or piece of a leaf, in its jaws, but so as to turn the left edge upwards along its back thus very efficiently masking its body. Is this a device to escape detection—for ants have many enemies—or is it, as some suggest, merely an accidental benefit, the leaf being borne along for the purpose of furnishing material for the cultivation of a fungus within its nest. But the matter is complicated by the fact that a totally unrelated insect—a tree-hopper—has developed a laterally compressed body terminating above in a leaf-like edge, and green in colour, so that it looks like a leaf. Has the tree-hopper mimicked the sauba ant, or is the case the other way about?

There is a psychological aspect of this theme of coloration and camouflage which is very puzzling. The behaviour of some of these creatures seems to indicate reasoning powers of a very high order. But this is an assumption that will not stand the test of criticism. Reasoning must obviously have memory of experience behind it, on which to shape conclusions.

Now, a caterpillar newly emerged from the egg can have no knowledge of the existence of the enemies by which it is surrounded, still less of the need for disguising its body in order to escape being eaten! Yet we have seen that the caterpillar of the emerald moth, as soon as it leaves the egg, proceeds to hook pieces of lichen on to its back. We may say, if we

M. Duncan

. Bastin

BUTTERFLY'S CUNNING TRICK: SEMI-INVISIBLE CRABS

When a grayling butterfly settles it does so upon the ground, and lies over to one side so that one wing shall be uppermost, harmonising with the background and not casting any shadow that might attract dangerous attention. The upper photograph shows a portion of a seashore containing several almost invisible crabs imitating their surroundings.

like, that they act instinctively, and we are set wondering *how* such instincts could have arisen, and the wonder is heightened by the recollection that their "instructive" actions are associated with special structures—like hooked hairs—without which their instructive responses would be useless. But here are the facts translate them as we will. We have to recognize an evolution of behaviour delicately adjusted to keep pace with the evolution of structure. The two must keep a very exact response to the demands of the conditions of the environment. If the wolf is to wear sheep's clothing successfully he must act like a sheep.

Very helpful in this connexion is the less complicated cases furnished by the sargasso-fish—one of the angler-fishes. It lives in the famous sargasso weed of the Gulf Stream, and in its coloration marvellously matches its surroundings, even developing small white spots to resemble the similar spots on the surrounding weeds due to small organisms known as bryozoa. Further than this, its pectoral, or breast fins have become transformed into claspers wherewith to clasp the weed, and so maintain an absolutely motionless poise without effort. This is essential; for it is invisible to its prospective prey only so long as it does not move.

Another dweller among seaweed is a large Australian species of sea-horse which has developed enormously long filaments of skin all over its body, thereby causing it to blend with the long leaves of the weed amid which it hides. But in this case the disguise is not so much to deceive its prey as to deceive its enemies. Here, again, unless the instinct to remain motionless is fully developed, the likeness of the body to its surroundings is unavailing. Structure and behaviour must always harmonise; they must be reciprocal.

And now let us pass to some other no less remarkable cases where peculiarities of structure and unusual behaviour have to keep pace with one another. These are furnished by caterpillars. Let us take the caterpillar of the puss-moth. Already coloured

Neville Kingston

so as to harmonise with the leaf on which it is resting it will, when alarmed, face the enemy, and retract the head so that it is surrounded, as in a frame, by the bulging walls of the body segment immediately behind. This brings into prominence a vivid red ring and two intensely black spots looking like eyes so that the bewildered enemy is suddenly confronted by a very weird and forbidding-looking face.

But this is not all. At the same time it turns the end of the body upwards over the back and thrusts out a pair of long, red, waving filaments! These filaments and their sheaths are worth a little attention, for they have come into being by transforming the last pair of hind-legs. Here, again, this special and very remarkable transformation of the hind-legs for a special purpose would be useless unless it was

E. Step

CRABS UTILISING SPONGE AND SEAWEED

Below is a crab that has decided to resemble the bottom of the sea more closely by attaching a growth of sponge to its back. Of course, when the crab moults it will have to find something else to disguise its new shell—some seaweed, for instance, as the crab in the top photograph is doing. Its claws are covered.

P. Webster

STONE CURLEW AND LESSER TERNS CLOSELY IMITATIVE OF THEIR SURROUNDINGS

As its name suggests the stone curlew has a partiality for pebbles, and particularly is this so when it wishes to lay and hatch its eggs. The eggs themselves and the young chicks both imitate so well the stones among which they lie that it is a hard matter to detect the presence of either. The lower photograph shows a young stone curlew in typical surroundings. Above are three young lesser terns, almost invisible among the stones and loose sand. If these little creatures start to run, the effect is as though the stones were moving.

A. H. Willford

A. B. Beattie

HOW A FAWN AND A LEVERET FADE INTO THE LANDSCAPE

When photographed this fawn (bottom) was only five days old, and at such an age protective colouring must come to the aid of weakness if the young stag's chances of surviving the dangers of wild life are to be anything but small. The upper photograph is of a month-old leveret crouching in the grass. Its fur is wonderfully matched to the background, although, of course, there is not a particle of green in the whole skin. Living in the open, the hare's ability for being invisible must of necessity be tremendous.

Mondiale

CHICKS AND EGGS OF PLOVER PRETENDING INVISIBILITY

Invisibility is made a fine art by young plovers. The lower photograph shows three of these birds squatting in the sand against which they look at a distance of a few yards like indistinguishable portions of the ground. Above is a Kentish plover's nest that is a depression in the shingle. Among the litter of stones and broken shells the three eggs are concealed as well by imitative marking as though they were buried. The Kentish plover gets its name from the fact that it was first noted as a separate species on the Kent coast.

A. H. Willford

P Webster

NESTING BIRDS ELUDING OBSERVATION: WOODCOCK AND RINGED PLOVER

A single watching eye is the most salient feature of this sitting woodcock which stretches itself straight out upon its nest, remains motionless and lets the mottling of its feathers settle the question of invisibility. Above we see a ringed plover in the act of pushing herself on to her eggs, which are laid in the open among the sands of some lonely shore. The eggs, of course, harmonise with their surroundings, and when the mother covers them they are still as well protected from observation.

S. Crook

accompanied at the crucial moment by the appropriate behaviour. No less remarkable is the caterpillar of the lobster moth. Coloured to resemble a dead leaf, its body is further transformed by angular prominences along the back, and exceptionally long fore-legs, while the hinder part of the body can be made to assume a curiously plump appearance when it takes up its terrifying attitude. This it does when alarmed. On such occasions the fore-part of the body is raised up and the long legs spread out, and set quivering as though in its frenzy it were striving to seize the intruder.

Finally, let me cite the singular case of a grasshopper which simulates a hissing snake. Professor Gregory tells us that during his exploration of East Africa he was strolling one evening out of the camp when he was startled by a hissing noise like that of a snake from a clump of grass. Springing back he pelted the grass with sand. But as no snake emerged he approached and peered into the clump, where he could just see a small green head which, whenever the hissing noise was made spread out like the head of a cobra. He tried to kill the creature by blows with his stick behind the head—and one of these knocked it over. Stooping to pick up the victim he found not a snake, but a grasshopper!

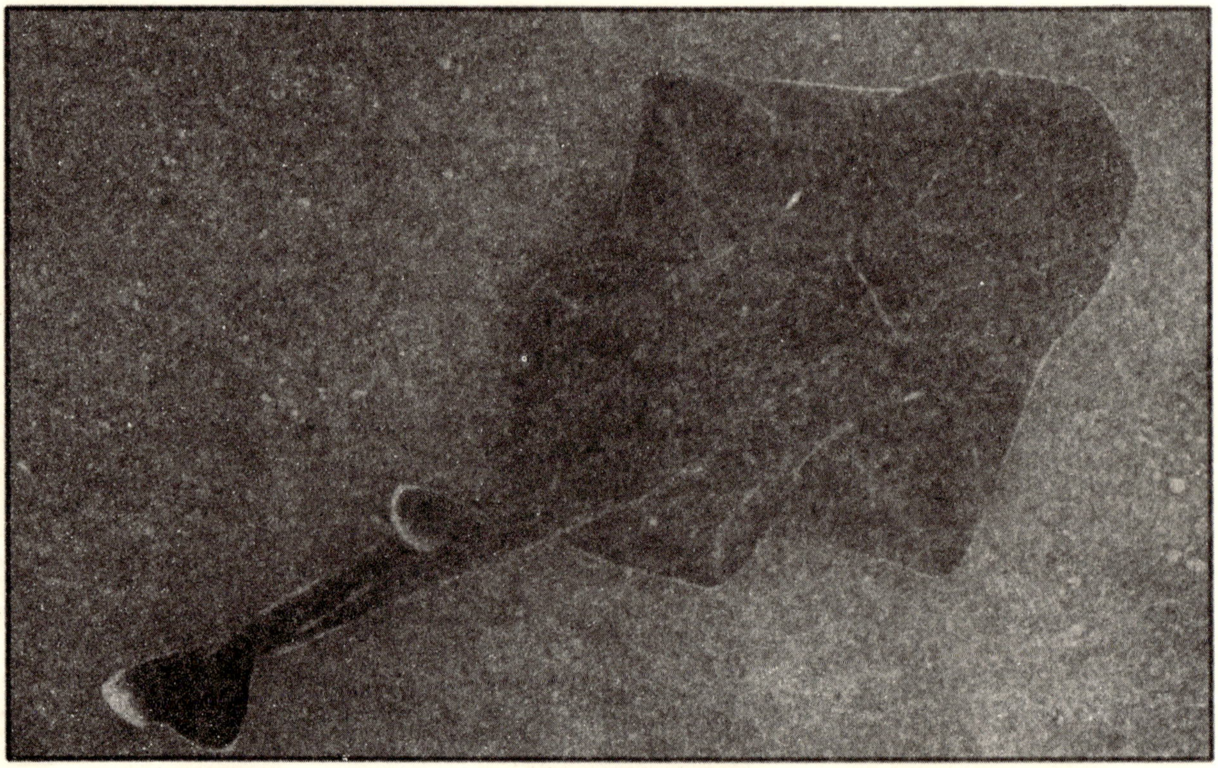

M. Duncan

MONK FISH SEEN BY THE CINEMATOGRAPH CAMERA: SPIDER SHAMMING DEATH

Fish that swim close to the bottom, such as the skates, plaice and halibuts, or the monk fish, of which last we have an example here, are light below and parti-coloured above, so that they harmonise with their surroundings. In this case the fish shows up much more than usual because, for photographic purposes, the light had to be much stronger and the water clearer than in natural conditions. Above is another case of make-believe—a spider (top photograph) journeying across a doorstep and (lower photograph) shamming death on being frightened

H. Bastin

J. J. Ward

WONDERFUL EXAMPLES OF MAKE-BELIEVE AMONG THE INSECTS

Like pieces of a broken branch when the wings are closed, the buff-tip moth (bottom left) has adopted a remarkable way of escaping the notice of enemies. The rose-leaf moth (bottom right) is still more ingenious, for when it folds its wings on a rose-leaf it resembles the droppings of a bird. Above are some cocoons of the fig-wort weevil (top left), the larvae of which are exactly like the seed capsules of the plant on which they live. A number of larvae of flata (top right) on a branch look rather like a piece of the plant "honesty."

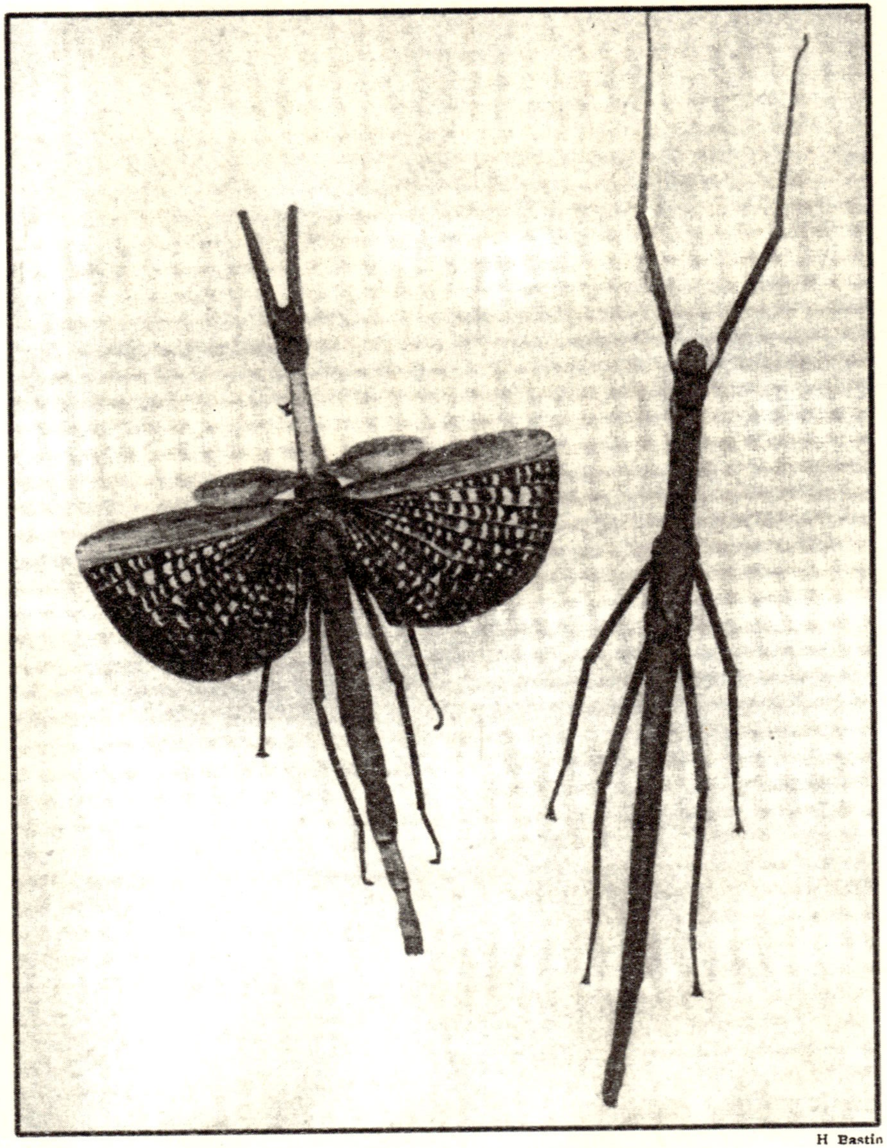

H Bastin

INSECTS EXTRAORDINARY: WEIRD-SHAPED CREATURES THAT LIVE AMONG LEAVES AND TWIGS

Members of the family of the Phasmidae, comprising the stick and leaf insects, are among the most queerly-shaped creatures that the world knows. The specimen on the left, for instance, resembles a grass-stem when its wings are closed and is said to behave like a ballet dancer when they are open. The right-hand photograph shows an other of these extraordinary-looking insects which so closely resemble the pieces of vegetation among which they live. These insects can, to some extent, replace lost or damaged limbs, though strangely enough, a lost leg is sometimes compensated for by a new antenna or vice versa. In some members of the family, also, the male is very rarely found and many generations appear which seem to be entirely composed of females.

Chapter LXXII

Insects of Strange Form and Habits

By F. Martin Duncan

Librarian to the Zoological Society of London

AMONG the many curious inhabitants of the insect world—and their name is legion—the praying insects, or soothsayers (*Mantidae*) have long attracted special attention not only from their strange appearance but from the almost world-wide reputation for piety and wisdom that they, quite undeservedly, enjoyed in olden times.

These strange creatures are found only in warm countries; they are mostly of large size, as insects go, and are usually coloured to blend harmoniously with their surroundings—being green, brown, or parti-coloured to match the foliage, grasses or flowering plants among which they habitually dwell. The whole "make-up" of the Mantis family is decidedly original. The fore-part of the body is unusually drawn out and has a curious resemblance to a long, giraffe-like neck, on the top of which is perched a queer-shaped head (usually triangular, but sometimes round or long) complete with two great, bulging eyes which the insect twists and turns about in all directions with a most absurd effect.

But the most singular part of a mantis's anatomy is the remarkable structure of its fore-legs. They are abnormally long and strong and the third joint, which we may call the thigh, is broad and flat and has a row of sharp, curving spines along the outer edge. The fourth joint, or shank, is also armed with spines and can be folded back upon the thigh like the blade of a clasp knife. The two rows of spines then interlock and form a veritable vice.

When resting in its characteristic attitude the mantis stands upon its four ordinary legs, raises the long, neck-like portion of its body, casts its eyes heavenwards, and uplifts its fore-limbs as if in supplication; and thus, motionless, the strange creature will remain for hours together. This devotional posture so impressed superstitious people in bygone days that the insect was looked upon with awe and wonder and credited with divers virtues and miraculous powers. The ancient Greeks believed it to be engaged in meditation on futurity and so bestowed the name "mantis"—that is, diviner—upon it. Pious folk declared that it was wrapt in prayer and to this day it is known in different parts of Southern Europe by such names as "prie-dieu," "prega-diou," and "fouva-dios," while in the East the Turks and Arabs assert that the mantis always "prays" with its face turned towards Mecca.

MANY quaint legends were invented about this gifted insect, and a very old naturalist gravely informs us: "So divine a creature is this esteemed, that if a childe aske the way to such a place, she will stretch out one of her feet and show him the right way, and seldom or never misse. Her tail is two-forked, armed with two bristly prickles; and as she resembleth those Diviners in the elevation of her hands, so also in likeness of motion; for they do not sport themselves as others do, nor leap, nor play, but, walking softly, she retains her modesty, and showes forth a kind of mature gravity."

But these hypocritical insects no longer impose upon us. Their true character now stands revealed, and they are known in reality to be the most savage, bloodthirsty little wretches possible—regular ogres of the insect world. Their saintly demeanour is merely a pose and "for ways that are dark and for tricks that are vain" a praying mantis could give points to Bret Harte's "Heathen Chinee," for when seemingly wrapped in profound meditation it is actually engaged in keeping a keen look out for something to kill and devour. The moment an unsuspecting insect alights near by, the mantis with slow and stealthy steps moves towards it; a long limb shoots out, and "click," the unhappy creature is caught and crushed in the diabolical leg-trap. After tearing its prey piecemeal with its strong jaws the mantis once again uplifts its paws, as if it were saying grace after meat, and patiently and piously awaits another victim.

BUT this is not all. Even when hunger is appeased the insatiable insect will continue to slay every small living thing that comes within its reach. It is quarrelsome, too, and whenever it happens to meet another of its own kind it at once begins to fight. The two attack each other furiously, lashing out viciously with their sabre-like fore-legs until one of the antagonists overcomes its opponent, not infrequently by slashing off its head. When the battle is over the victor forthwith proceeds to make a meal of the vanquished and if, as is often the case, a male and female have been fighting it is usually the male, who is the smaller of the two, that furnishes the feast.

Little in the way of food comes amiss to the mantis; it preys upon flies, bees, caterpillars, grasshoppers, and all soft-bodied insects; tough-skinned creatures such as beetles are killed but usually soon discarded, while certain large South American species are said to capture and devour such prey as small frogs, lizards, and even birds.

There are probably quite six hundred different species of mantis distributed over the warmer parts of the world. The insects are from one to six inches in length. Many have brightly-coloured hind-wings, resplendent in red, mauve, pink and blue, but when the insects are at rest they are concealed beneath the fore-wings, which are thicker in texture and usually leaf-like. Some species, in addition, have leaf-like expansions on their legs, while others are so flower-like that even butterflies are deceived by them.

F. W. Bond

The variety of forms assumed by these deceptive insects is amazing. Not only are leaves, lichens and flowers simulated by different members of the mantis family, but certain species mimic other types of insects. One mantis so successfully counterfeits a harmless stick-insect that its disguise would be perfect if its raptorial legs did not betray its real nature.

The stick-insects (*Phasmidae*) are totally different in character from the rapacious *Mantidae*, although in appearance they are, perhaps, even more extraordinary. Their popular names—stick-insects, or walking-sticks—describe them fairly well; for although different species show a remarkable variety of form and colour the majority have a distinct resemblance to sticks, twigs or plant stems. They are a feeble, timid folk, whose chief anxiety in life, apparently, is to escape observation. In this they are eminently successful, as in their form, colour, and the attitudes they assume, they conform to the natural objects among which they dwell in so marvellous a degree that only a keen and practised eye is at all able to detect them.

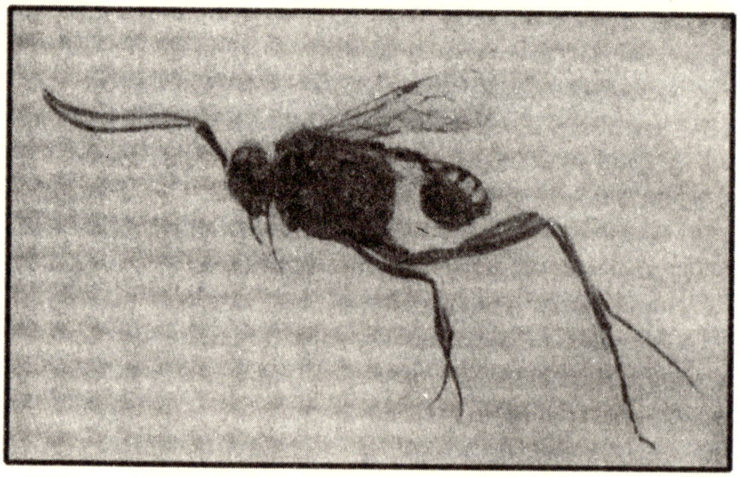

H. Bastin

ECCENTRICITIES OF SHAPE

Below is an amazingly shaped wasp-like parasite with an extraordinarily small hind body and an unusually placed waist attachment. Above is one of the more bizarre of the rapacious insects belonging to the mantis family. Not only is its attitude grotesque, but its bodily form also.

One species, for example, that lives among long grasses is almost a replica of a grass-stalk. Its long, cylindrical body is identical in thickness and its colour is green when the grass is fresh and turns correspondingly yellow or brown as it fades. Its favourite habit is to rest with its feet clinging to a tall grass-stalk, its almost thread-like legs—that is to say the two hinder pairs—pressed closely against its sides, while the fore-legs, which are very long, are stretched straight out in front in such a way that the insect's head, with its eyes and antennae, is hidden from view between them. In this attitude it will remain, perfectly motionless, all day; but when the light fades the "walking-stick" comes to life and begins to move slowly and laboriously over the grasses, its long body swaying from side to side on its inadequate legs, which seem too feeble to support its weight. Having, probably, fasted all day it is extremely hungry and it spends its short period of activity in feeding vigorously on the surrounding vegetation..

Some of these insects, (frequenting damp forests in tropical America, and the Malay isles) to make their disguise more convincing, have the most extraordinary excrescences dotted all over them, making them look like moss-covered twigs; others have leaf-like processes sprouting from their bodies and legs; others again are covered with prickly spines and look exactly like the thorny stems of the bushes on which they make it their business to live.

Many forms are wingless, but others possess delicate and beautifully-coloured wings which, however, are not conspicuous when the insects rest in their normal quiescent state. They fold up like a fan, lying flat and close against the sides of the insects without destroying the lineal outline of their stick-like bodies. Some species are quite small, barely an inch long, but others grow to a length of eight or ten inches, though even then they are hardly thicker than knitting needles, while a gigantic Australian walking-stick (*Acrophylla titan*) is a foot long and as thick as a man's finger. Being strictly vegetable feeders the larger kinds sometimes seriously damage forest trees. A curious fact in connexion with these insects is that they possess the power of reproducing lost limbs. The new limbs, however, are never quite as large as the original ones, and what is even more curious is that they do not always grow true to type. For instance, an absent-minded stick-insect will occasionally replace a lost antenna with a new leg, or vice versa.

Equally anxious to pass unnoticed amidst the teeming life of the tropical forests are the walking-leaf insects. They are nearly related to the stick-insects, but have devloped the family trait of dissimulation on different lines. Instead of being long and thin they are broad and flat, and in shape and colour approximate so closely to the foliage of the trees that form their natural habitat that they are practically invisible unless they destroy the illusion by moving.

H. Bastin

INSECTS EQUIPPED WITH BEAKS

From South America, home of so many strange forms of life, comes an insect whose official name is Rhinastus. Not only has it a long thick beak, but also strange processes on the ends of its legs. Above is a nut weevil, with its thin, curved beak, nearly as long as its body.

The fore-wings of a walking-leaf are typically leaf-like and lie flat over the depressed body of the insect, while the legs appear as small leaflets, or bits of leaves, projecting from the sides. To heighten the effect the head, together with the first segment of the body, represents a short swollen leaf-stalk. So perfect is the impersonation that in the East Indies, where the creatures abound, the natives firmly believe that they actually are leaves which have been miraculously transformed into insects.

That the leaf-like nature of these peculiar creatures is more than skin deep is suggested by the fact that, although strictly vegetarian in their diet, if two walking-leaves are shut up together without their natural food, they will satisfy their hunger by nibbling little pieces out of each other's wings and legs. It is worthy of notice, too, that the eggs of both stick and leaf-insects are so extraordinarily like seeds in appearance that even botanists have, at first sight, been entirely deceived as to their true nature.

Another insect that is worthy of notice on account of its strange mode of life, and the cunning way in which it overcomes its natural deficiencies, is the ant-lion, or rather, we should say, the larva of the ant-lion, for the perfect insect is in no way remarkable in its habits. On emerging from the egg the young ant-lion is a veritable "ugly duckling," in every way the antithesis of the graceful, gauzy-winged insect it will eventually become. It has a squat, bulky body, rather like a compressed egg in shape, beset on each side with numerous tufts of black hairs and surmounted by a large, flat head armed with a formidable pair of sickle-shaped jaws. Its legs are too feeble adequately to support its heavy bloated body, and as the hind pair are permanently fixed against its sides the insect is unable to move in a forward direction, but is condemned throughout the whole of its larval life to shuffle slowly backwards. Now, the ant-lion is a carnivorous creature, and if it were obliged to rely solely upon its activities it would fare very badly—for a retrograde mode of progress is not well suited to the capture of lively insects. So the wily creature resorts to strategy in order to procure a sufficiency of animal food.

Choosing a site where the soil is dry and sandy the ant-lion walks slowly backwards, describing a circle on the ground. As it goes it shovels the loose sand on to its flat head and with a sudden jerk flings it outside the circle. Round and round the odd little creature plods, working in a spiral until it has succeeded in excavating a funnel-shaped pit two inches deep and about three in diameter. Then, its task accomplished, it proceeds to bury itself in the sand at the bottom of the pit, leaving only its jaws and the front part of its head exposed to view, and patiently awaits the reward of its labours.

It seldom has long to wait. The chosen locality is always swarming with busy ants, fussy little beetles and such "small deer," and many an inquisitive insect pauses on its journeyings to and fro to peer down

H. Bastin

Martin Duncan

PRAYING MANTIS, GIANT LANTERN FLY AND A STALK-EYED FLY

The word "mantis" is a Greek one and means soothsayer, for this insect of strange form was thought of old to be endowed with strange powers. The bottom left and top right-hand photographs show two of these creatures. The mantis was once believed to be a benevolent guider of lost children, pointing the road for them with outstretched leg should a child ask the way. In reality, the mantis is the fiercest insect in existence, so far as other insects are concerned, and kills not only to feed but for the mere lust of killing. The bottom right-hand photograph shows us one of the giant lantern flies of Brazil, while in the top left-hand photograph is a fly from south-eastern Africa, which has eyes on stalks, and the stalks nearly as long as the outspread wings.

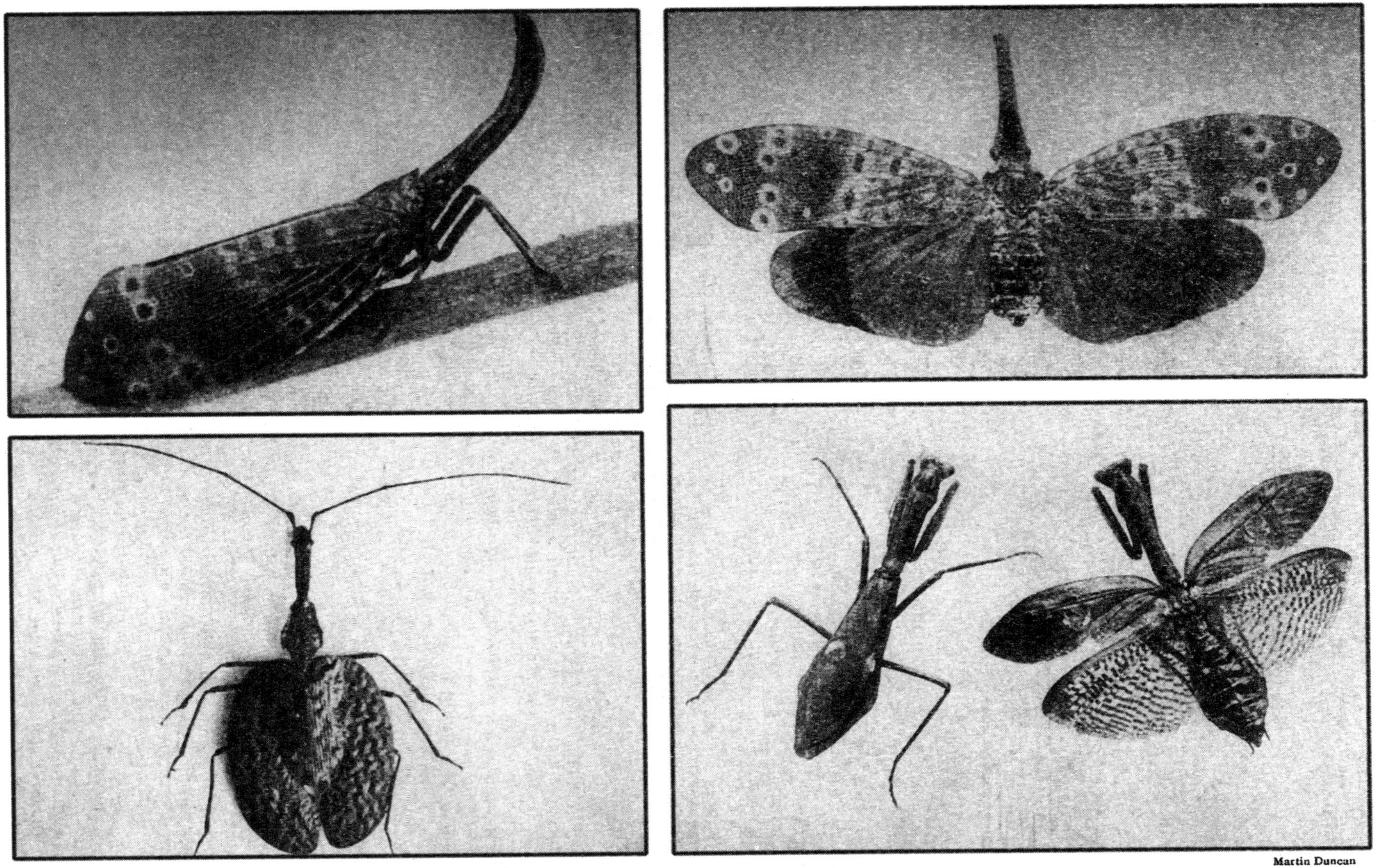

Martin Duncan

FOUR REMARKABLY-SHAPED INSECTS FROM THE TROPICAL FORESTS

Among many remarkable beetles is the "fiddler," or violin beetle (bottom left) found in China and Java. Its body, shaped something like the outline of the back of a violin, is so flat that the beetle can squeeze itself into very narrow crevices, and this it must do since it makes its home under the bark of fallen trees. Another extraordinary insect is the lantern-fly (top photographs), which received its name because, in some way, the idea arose that its large, ungainly proboscis was illuminated at night. This idea has been proved erroneous, but the purpose of such a snout remains a mystery. This specimen comes from Tongking and is seen with its wings open and closed. Below (bottom right) is a mantis, also with wings open and closed.

Insects of Strange Form

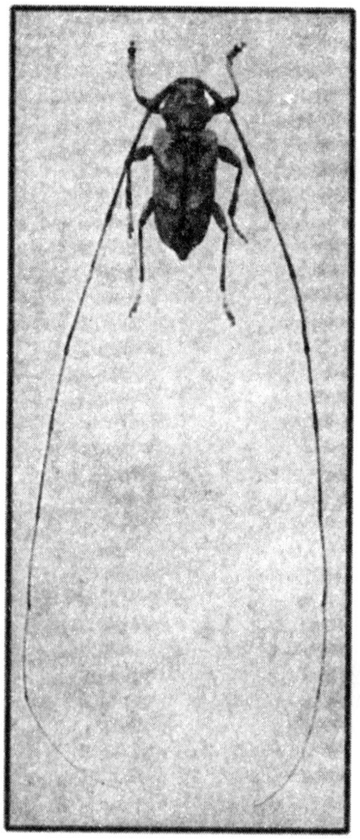

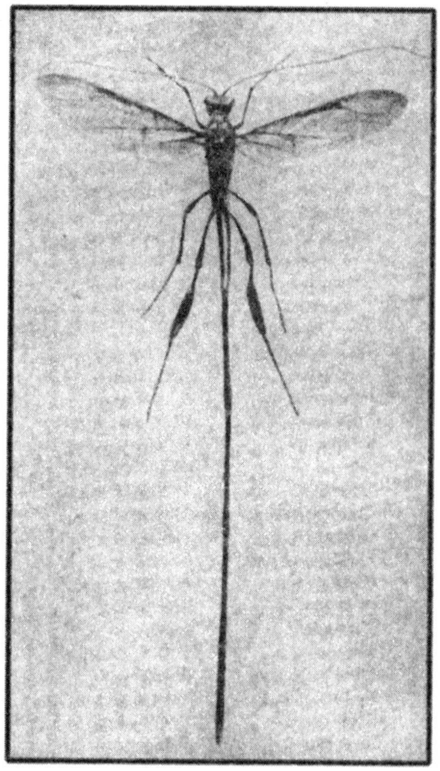

H. Bastin

EXAGGERATED FORMS OF INSECT BODIES

Here is a strange species of British origin, the timberman beetle (left), which is mostly found in Scotland, where it lives in fir trees. The remarkable antennae are very considerably longer than the insect's body. The centre photograph shows us the male of the long-headed weevil. The strange prolongation of the fore part of the male is well seen here. The right-hand photograph is of a weird parasitic, wasp-like insect with a very long and greatly attenuated hind body.

into the pit. This is its undoing; the loose sand on the bank gives way under its feet and the luckless insect begins to slip down the sloping sides of the trap. Its struggles only make matters worse; it rolls helpless down the treacherous incline into the jaws of the ant-lion, who hastens the denouement by frantically scooping up the sand with its head and flinging it in showers over the poor thing. Whether this manoeuvre is intentional or merely the outcome of the ant-lion's excitement at seeing its dinner arriving it is impossible to say. In any case, the effect is the same, and the savage little creature after sucking the juices of its victim again brings its useful head into play, and tosses the empty skin outside the boundary of the pit.

When full grown the ant-lion's appetite fails, and it constructs a globular cocoon of grains of sand bound together with fine silken threads produced from a slender spinneret at the end of its body. Within this cocoon the insect rests until its metamorphosis is complete, when it emerges transformed into a slim-bodied, gauzy-winged creature not unlike a dragonfly.

BUTTERFLIES and moths, on the whole, show less originality in form and behaviour than other types of insects. Many of the lepidoptera, of course, are protectively coloured, and of these the well-known Indian leaf-butterfly (*Kallima inactus*) that mimics a faded leaf in the dry tropical forests, is a striking example, but any departure from the recognized customs of their kind is usually confined to their caterpillars. To feed and grow is the sole purpose of their larval existence, and to escape the keen eyes of insectivorous creatures their chief anxiety. This the majority achieve by protective coloration, though certain species—such as the larvae of the puss moth and lobster moth—bravely defy their enemies by assuming aggressive attitudes to scare them off. The caterpillars of the Psyche moths, however, carry the art of self-protection a step further by fashioning for themselves the most ingenious homes or garments wherein they may hide their plump and tempting little bodies completely.

The Psyche moths (*Psychidse*) have a wide distribution, and are known in various parts of the world as "basket worms," "case-moths," "house-builders" or "sack-bearers." Each species in building its habitation uses the material and follows the traditional design peculiar to its own particular branch of the family. The caterpillar of one little Psyche moth may often be seen wandering over heaths and grasses attired in a voluminous garment strangely reminiscent of an old-fashioned flounced skirt. This odd contrivance is made up of several rows of tiny bits of leaves arranged one above the other and finished off round the top with a row or two of fine fragments from sprigs of broom or heather. The whole affair, which is about an inch long, is held

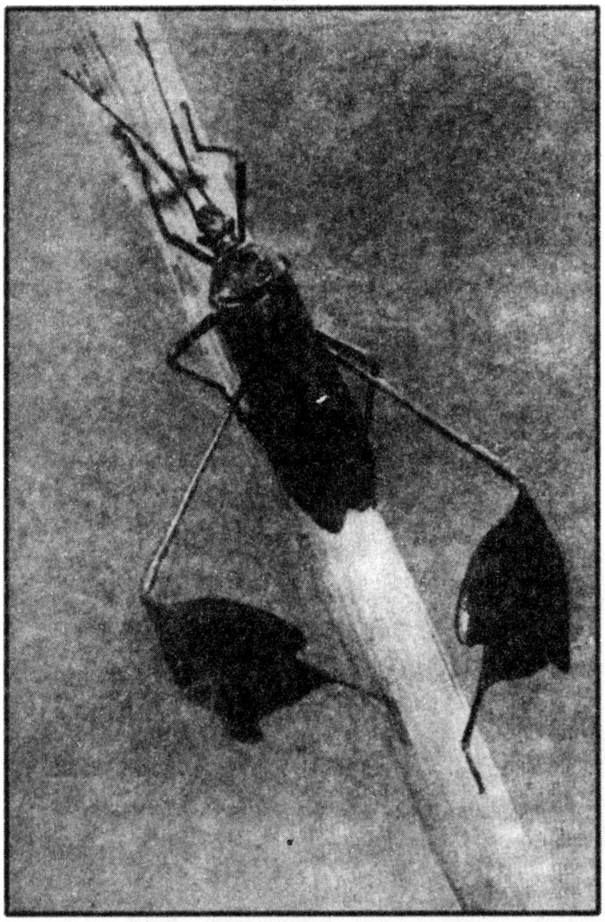

H Bastin

CAMOUFLAGED INSECT FORMS OF THREE CONTINENTS

In the great island of Madagascar there lives a kind of beetle whose name is Lithnius, and its body is formed to represent a growth of lichen. When settled upon a branch covered with real lichen (bottom) the deception is perfect. The kind of mantis called Gongylus (top right), which comes from Southern India, is coloured like a flower and, when it hangs head downward as is its habit, poses as an orchid to ensnare its prey. South America produces a bug (top left) which develops leaf-like enlargements of its legs. Its official name is Diactor.

Martin Duncan

HOW THE MURDEROUS MANTIS IN ITS FAMOUS ATTITUDE OF PRAYER APPEARS TO ITS HAPLESS VICTIMS

As soon as some insect alights near a mantis this formidable creature which, while it remained motionless in its "praying" attitude was virtually invisible, begins to crawl very cautiously towards its intended meal. When within striking distance a surprisingly long foreleg shoots out, seizes the wretched insect and then the deadly killing apparatus on the third joint of the foreleg comes into action. As we see here, it is armed with a saw edge and the fourth joint, similarly equipped, folds back on to it like the closing blade of a penknife. The mantis then tears its prey to pieces with its jaws and, having devoured the meal at leisure, resumes its famous "praying" attitude in preparation for the seizing and rending of the next insect to come within reach.

Martin Duncan

H Bastin

HORNED SCAVENGER BEETLE AND LEAF-LIKE MANTIS

Brazil produces a scavenger beetle (bottom) armed with a single up-curving horn and armoured like some weird war-machine of an imaginative novelist's fancy. This is one of the very strangest forms in the beetle world. The upper photograph shows us yet another form of mantis which should be compared with those seen in previous pages of this chapter. It has leaf-shaped prothorax and elytra, or wing cases, and thus is able to disguise itself very effectually from the insects on which it preys

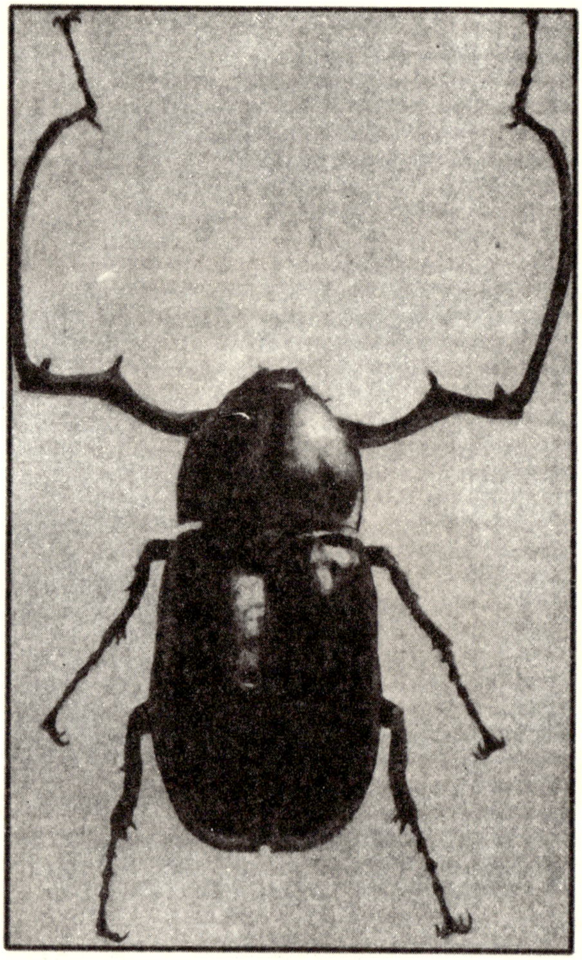

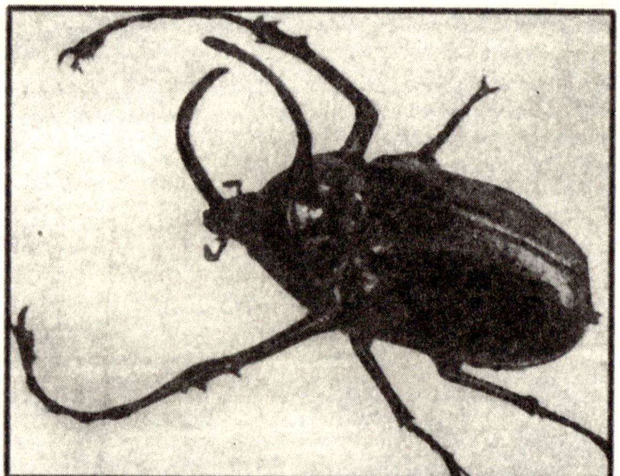

E. N A

SAWYER AND PAINTER BEETLES

Below we see a weirdly fashioned sawyer beetle with the two serrated horns on its head and its widely spread front legs. Above is a long-armed painter beetle, which has an enormous span to its forelegs and a heavy, thick-set body.

together with threads of silk, and has a soft, silken lining. In its perambulations in search of food the Psyche trails its case behind it wherever it goes, holding firmly to the inner lining with its hooked hind-feet so that only its head and scaly fore-legs are to be seen by a keen observer protruding from underneath the little bundles of leaves.

Another large foreign species constructs a long, tapering case of fine grass blades all cut to a given length and most neatly and symmetrically arranged. Chips of wood, fragments of straw, little bits of stick, lichens or mosses, and particles of earth are used by others to form their habitations, and one Asiatic Psyche makes a beautiful little case entirely of silk in the shape of a small, flat, snail shell.

From time to time, as the larva grows, it enlarges its case by adding to it a fresh row of its favourite material; and should an accident befall it the industrious insect at once sets to work to repair the damage by mending the rent, or fitting in a neat patch. When ready to undergo its transformation the larva carefully closes the larger end of its tube, and having thus secured itself against intrusion, changes into a chrysalis.

So, still protected by its handiwork, the little creature rests until its metamorphosis is completed. Then, if the insect is a male, a small delicate moth emerges from the case and flies away to enjoy a brief period of liberty in the open air. But if the moth is a female she does not leave her prison. She is wingless, and sometimes legless as well, in fact, little more than an animated egg-bag; she lays a quantity of eggs and, her mission in life accomplished, she at last crawls forth only to die.

The old case now becomes a nursery, but the baby caterpillars do not inhabit it for long. Almost immediately after they are hatched the tiny creatures make their escape and each one sets to work without delay to make a little case for itself exactly as its parents did before it.

In most cases, no matter how bizarre in appearance, or odd in its ways an insect may be, its eccentricities have, obviously, some direct bearing on its own peculiar mode of life; but there are some instances in which, to the ordinary human mind, these idiosyncrasies appear to have no special significance—though doubtless if we could see them from an insect's point of view we should find they were important in one way or another.

Of what use, for example, can the huge, misshapen proboscis of the lantern-fly (*Fulgora laternaria*) be to the insect? It was at one time supposed to be luminous and to shine forth like a lantern at night—but this idea was proved to be erroneous. The use of the proboscis is still a mystery.

Among the vast number of extraordinary beetles to be found all over the world none is perhaps more remarkable in shape than the singular violin beetle (*Mormolyce phyllodes*), or the "fiddler" as it is sometimes called. It is a native of Java and China, where it lives in the forests hiding under the bark of fallen trees. It is so surprisingly flat and thin that it can squeeze its body into the narrowest crevice, while its general shape and highly-polished, dark brown colour give to the insect the strange likeness to the musical instrument from which it takes its name.

Chapter LXXIII

Horns of Beauty and Horns of Strength

By Frank Finn

Author of "Wild Animals of Yesterday"

THERE is a well-known old story about a French student who tried to play a practical joke on Cuvier by dressing up as the devil, with tail, cloven hoofs, and horned and fanged mask, and arousing him from sleep with the words, "Cuvier, I've come to devour you!" To this impertinence the great scientist replied, "Horns and cloven hoofs with carnivorous dentition! An impossible combination—you can't do it!"

It is not quite so easy to be positive nowadays about what is a possible combination of characters. Cuvier would have been astonished to know of such a beast as has since been discovered, with paws and the general structure of a carnivore and herbivorous teeth—the opposite combination. But the story will serve to introduce two important points in connection with horns—their correlation with hoofs, especially with the two mutually-adjusted hoofs we know as the cloven hoof, and the paradoxes which really are often associated with them.

We may take as the first of these the fact that horns are very rarely wholly made up of the substance we call horn, and that the animals which are provided with such are not successful in the struggle for existence. For rhinoceroses are the only beasts whose horns are solid horn all through, and rhinoceroses are some of Nature's failures. All the living kinds can be counted on the fingers of one hand, and man has met but one other in life—the extinct woolly rhinoceros of the North, the portrait of which he drew in his old Stone Age days.

Rhinoceroses have always been rare in captivity, and one curious point about their behaviour in this condition may be noted—the fact that the captive Indian rhinoceros has a trick of grinding its horn down to a flat plate against any hard surface, while the common African rhinoceros leaves it alone. The great so-called white rhinoceros has never been exhibited alive; this has the largest horn of all of them, in the case of the first of its two horns.

Asiatic rhinoceroses have no doubt been much reduced owing to the idea that the horn was a poison-indicator when fashioned into a cup which, it was believed, would split if poison were poured in; for such an idea would naturally induce a brisk demand, and it probably arose from the fact that the solid horn was a type of what is true and genuine.

W. S. Berridge

HORNS OF THE MUSK OX

With the musk ox the horns are curiously flattened against the head and almost look as though the fibres of which they are made had been first brushed hard and then had set. This formation turns the head into a powerful battering-ram at need.

The Indian rhinoceros was the original of the so-called unicorn, but to make this mythical beast there has evidently been some confusion with a long horned antelope, and it has besides been given a horn, from a widely-different beast, which is not a horn at all, but the long, straight, spirally-grooved tusk of the small whale we know as the narhwal. That two quite different creatures can be mixed up in the mind was proved to the writer when he heard a woman visitor to the pond by the Palm House at Kew speak of a mandarin drake as a gold pheasant!

The rhinoceroses are the only hoofed animals with actual horns outside the cloven-hoofed section of the group, but it is a curious fact that in rare cases the horse, presumably as a legacy from the past, shows a tendency to develop them, two bony bosses appearing on the forehead under the skin.

this family display horns which, in spite of great variety, agree in their plan of structure. They are plated articles, consisting of a core of bone covered by a sheath of horn, which may be either smooth or marked by raised rings or lumps. Many of these horns are exceedingly beautiful, and it is difficult to say which is the finer, the backward scimitar sweep of the sable antelope and the ibex, or the wide-flung corkscrew spiral of the koodoo and the markhor, to say nothing of the massively simple open curves of cattle horns or the bold circular turns of the giant sheep or argali.

Other patterns verge on the grotesque, like the zigzags of the hartebeests and the vicious-looking pot-hooks of the musk-ox, which is one of the capricorn group, despite its appearance.

But this development, if it proceeded further, would result in the formation of a different type of horn from that of the rhinoceroses. It has done so in the giraffe, in which conspicuous though short horns exist, which are always covered with skin. Here not only two are found, but a central one lower down and less developed may occur, and even two in addition to this. In the giraffe's more primitive relative, the okapi, the female has short, skin-covered stumps, but in the male a cap of bare bone projects from the top of these, thus foreshadowing the state of affairs in the deer family.

PROVIDING the most numerous and varied of the horn-bearers is the bovine family, which includes not only the true bovines or oxen, but the sheep and goats, the capricorns or goat-antelopes, and the vast and varied assemblage lumped together as antelopes, most of which look like deer, though some, like the gnus, are horse-like, and others, like the elands, remind one of oxen. All of

CAPTIVE INDIAN AND AFRICAN RHINOCEROSES

When in captivity the Indian rhinoceros (bottom) has the strange habit of rubbing its snout against any hard surface so that the horn gradually becomes worn down to a flat plate as seen here. The African rhinoceros (top) has two horns. Of the hoofed animals whose hoofs are not cloven, the rhinoceros is the only one possessing a horn.

HUGE PALMATED ANTLERS THAT ARE BORNE BY THE MOOSE

Enormously heavy plates of horn armed with strong spikes make up the armament of the moose. It was this type of horn that was possessed by the extinct giant fallow deer of Ireland, whose remains are sometimes dug up. These have yielded some of the finest antlers known and have been found measuring four feet across with palms as large as the seat of a big chair. The moose, possessing this type of heavy antlers, has them perfectly set on the head so that the balance is just right

W. S. Berridge

HEAD OF BLANDFORD'S SHEEP FROM BALUCHISTAN

This is a variety of the Urial, or Punjab wild sheep, and is distinguished by its very wide-spreading horns. These are prominently ridged and their curves make, in the case seen here, an arc that is considerably more than a semicircle. The bold simplicity of their sweep is magnificent, and the well-marked ridges add to the appearance finely. No wonder that so much trouble is taken over the obtaining of a fine pair like this as a sportsman's trophy. Blandford's sheep is found in the wilds of Baluchistan.

In the musk-ox and the African buffalo the horns expand and often meet at their bases, forming a strong shield for the forehead. In the Indian buffalo the horns are triangular in section, and attain a great length in some cases, especially in the cow, which has longer if less massive weapons than the bull. A horn of this species is known which actually exceeds six feet six inches in length.

As a general rule, the males of these hollow-horned ruminants bear larger horns than the females, but they may be hornless altogether, in which case, if of the usual deer-like shape, they are difficult to distinguish from female deer, unless one knows the particular species of both families.

THE power these arms confer on their wearers is very great. Even the lion has been found dead along with his prey, when this has been the oryx antelope which has impaled him on its long, bayonet-like horns, and the sable antelope at bay has been known to inflict casualties on dogs at the almost incredible rate of eight to the minute.

This beautiful creature is also so quick with its weapons that it can even parry flung assegais with them ; yet it is not so ready to fight as its shorter-horned but larger relative, the roan antelope, which is apt to lose its temper and charge if driven too hard. On the other hand the koodoo is said to have no idea of making use of its fine horns in defence, and the writer has known the markhor, also gifted with grand spiral weapons, knocked out by a mouflon ram far less than his size, though the markhor's relative, the tame goat, can certainly hold his own with the tame ram.

More, however, depends on the courage of the individual species than on its armament ; the African bush-buck, hardly larger than a goat, and short-horned, is a notorious fighter, and the Himalayan serow, a clumsy-looking capricorn with horns no longer than his donkey-like ears, has been found dead along with a whole pack of dholes—the terrible "red dogs" against which not only other horned animals, but the leopard, the bear, or even the tiger, cannot stand. The canine pack, indeed, seem to inspire a peculiar and unnerving horror in their prey ; the cats stalk and spring, and kill or miss, or, as we have seen, may be killed themselves ; but to be run to bay by the long relentless pursuit of numbers is an ordeal which requires the highest courage in the victim if it is to hold its own at the final encounter.

SHEEP WHOSE GREAT HORNS MAKE THEIR BODIES SEEM SLENDER

Of all sheep the argalis are the largest and they are found in Siberia, Mongolia and Tibet. We have a fine specimen here (bottom) with its curving, wrinkled horns which make the body seem almost slender and rather deer-like by contrast. The upper photograph is of a Barbary sheep, a native of the mountains of North Africa. Its distinctive feature is a kind of beard which grows on the forequarters and hangs down well below the knees. The horns are smoother than those of the argali, lacking the prominent ridges of the latter.

W. S. Berridge.

HOOKED HORNS OF THE CHAMOIS AND BRANCHED REINDEER ANTLERS

The chamois, or izard (left), of the higher and more isolated mountains of Europe, is distinguished by the structure of its horns from the other goat-like antelopes. The horns are set close together and rise almost vertically from the head till they suddenly bend back to form a hook. The chamois herds keep to the high woods in the winter and feed largely on lichens and mosses, while in the summer they live just below the snow line. Persecution has sadly reduced their numbers. The reindeer (right), with antlers like coral branches, is distinguished in its armament from the other deer by the fact that the brow-tines are palmated and also in that both sexes have antlers.

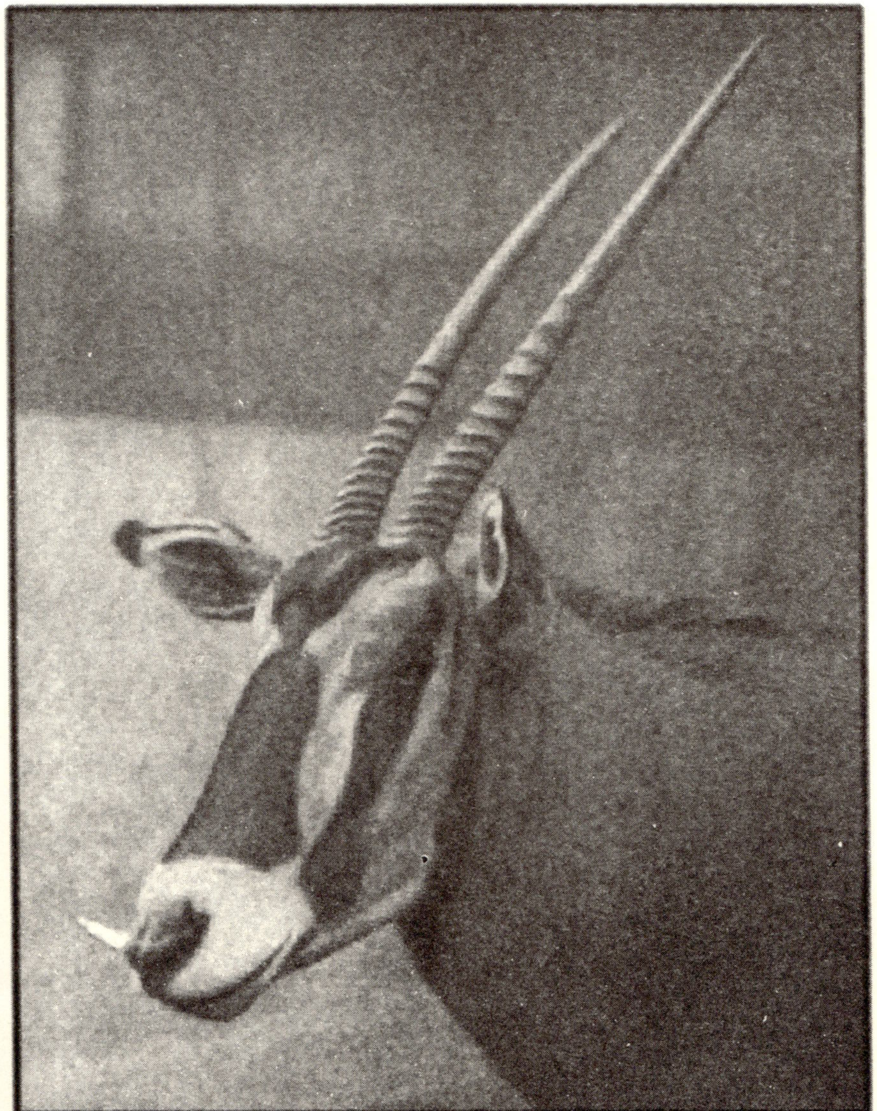

W. S. Berridge

TALL HORNS, RIDGED AND TAPERED, OF THE SABLE ANTELOPE AND BEISA ORYX

A good head of a sable antelope is one of the sportsman's dearest prizes. The horns are particularly fine and sometimes grow to a length of four feet or even more. They are thick and boldly ridged, and have a fine backward sweep to them. The head is black with white markings and there is a thick and handsome mane. Another very handsome head is provided by the beisa oryx (right). It will be noticed that the horns are ridged and that the ridges become more widely spaced towards the top and disappear altogether about half way up. Beisa oryx are found in large herds, and the sight of hundreds of their long, tapering horns all sloping one way makes as brave a show as a company of pikes in a historical picture.

W. S. Berridge

HORNS OF UNUSUAL DESIGN IN VARIOUS SHEEP

Nepal includes among its animals a strangely horned sheep called the unicorn sheep (bottom left). In this species the ram has its horns joined at the base and united for rather more than half their length. The horns then curve away from each other slightly, only to bend inwards again until, finally, they actually cross at the tips. In the barwell sheep, both of the black-faced (top left) and white-faced (top right) varieties, the horns are set far back on the head and sweep forward The fourth photograph (bottom right) *shows a sheep with quadruple* horns.

Horns of Beauty

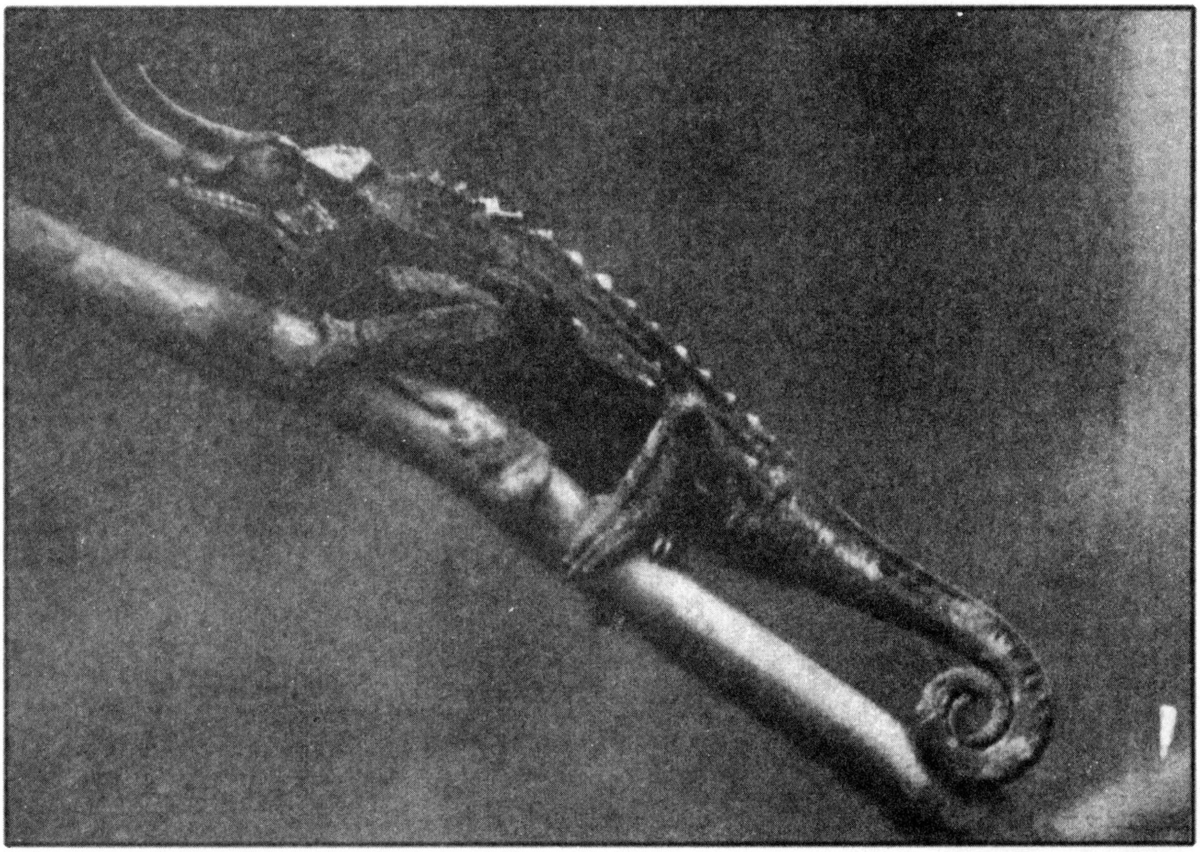

W. S. Berridge

HORNED CHAMELEON AND ITS STRANGE HEAD

So far we have been concerned with horns which were made for defensive purposes. But here is an inoffensive and rather defenceless little creature relying on its colouring to escape observation rather than on any powers of retaliation when attacked. The horns are peculiar to the male and seem to be purely for show since they could be of no service as weapons and they are not known to serve any particular function. Horned chameleons are found in Central Africa. Some species possess three of these weird ornaments.

Thus it is that canines clear out a district in which cats live without seriously incommoding their prey.

That the large oxen should be formidable is only to be expected from their size; they would be dangerous foes if they had the merest stumps of horns; but they differ greatly in temperament, the buffaloes, both African and Indian, being far fiercer than most. Indeed, even tame Indian buffaloes will attack the tiger unprovoked when in a herd, and are often used for driving a wounded one out of cover.

They dislike other animals, if strangers, besides this enemy, and perhaps, when they are attacked and sometimes killed by the gaur, or so-called Indian bison, they may have been "asking for trouble"; for the gaur, as far as man is concerned, is usually considered a harmless animal, whereas the wild bull buffalo is just the reverse.

There is, however, always a likelihood of warfare between two related species, and the writer once just missed seeing a really interesting bull-fight at the Zoo, because the fence between them held, though shaking perilously at each charge, when an American bull bison was trying to butt his way through to the English wild bull on the other side.

Many years ago one of these park bulls gave an example of the vicissitudes to which the wild warrior is exposed in nature; the second and third bulls conspired to depose their monarch, and handled him so roughly that he retired to nurse his injuries awhile; afterwards, however, he came out, killed one of his assailants, and regained his position. In a case observed in tame buffaloes in India, the master bull had to face three adversaries, and thenceforth was a very crestfallen and subdued beast.

Cases like this show that individual armature is not everything, and that cunning plays its part as well as strength even in contests between members of the same species.

The bovine horned animals mostly belong to the old world, but America has a few—the musk-ox, the Rocky Mountain "goat," one of the goat-antelope group, and in particular the American bison, which has achieved a greater reputation as a fool than as a fighter. Now and then, however, an animal with an extra good idea of looking after himself has occurred, and a redskin brave has had to gallop for miles to escape the determined pursuit of such an one; while a really beautiful instance has been recorded of a party of old bulls protecting a small

PERSIAN AND EUROPEAN RED DEER WITH MASSED ANTLERS

There exists this strange paradox about deer, that a deer's "horns," or antlers, are not made of horn, but of bone. Also, the deer with h antlers of many branches lives usually in forests where there is every kind of obstruction and impediment to progress for an animal so equippe It is the open horned varieties, antelope and oxen, which inhabit open country as a rule. In Northern and Central Europe the red deer seer to have deteriorated in size, but in Asia representatives of the red deer family are large and fine like this example from Persia (bottom

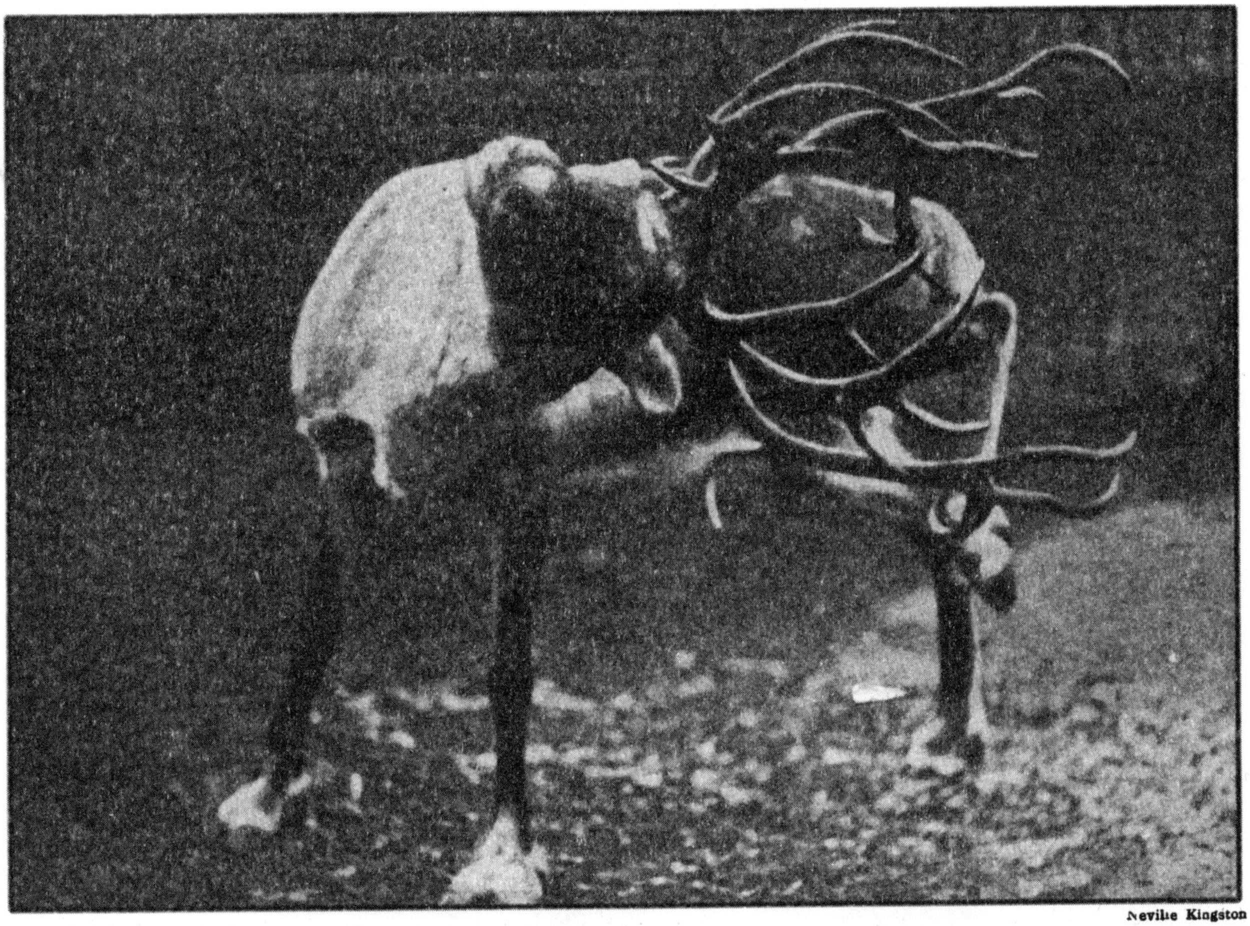

Neville Kingston

A HANDY USE FOR THE FAR-REACHING ANTLERS OF A REINDEER

Many animals can scratch their heads with their hind legs, but few have so fine an implement for relieving irritation in the hind leg itself as this reindeer of the London Zoological Gardens. This particular animal, which had formerly been quiet enough, suddenly developed a temper, attacked its keeper and smashed his wrist. The reindeer is one of the three kinds of deer which have developed the palmated or broad-ended type of horn, the other two being the moose and the fallow deer. Both sexes of reindeer bear horns.

weak calf from a band of wolves. The American bison is formidable from his great weight in front and from his speed, remarkable for an ox, but his horns are but poorly developed.

AMERICA has one beast called antelope, and a typical antelope in general appearance and in speed, which has the distinction of carrying a quite unique type of horn. The weapons are set far forward, over the eyes, in fact, and are hooked backwards like those of the chamois of Europe. But they resemble those of deer in being branched, though they have but one prong, far up on the front, and in being shed yearly, although they are composed of horn mounted on a bony core like the permanent horns of the family we have been considering. The core, however, does not extend into the prong and hook, and the mode of shedding is quite different from what is found in deer, the old horn being forced off by a hairy growth beneath it, while those of deer break off, leaving raw wounds. The prong-buck, as it is often called, has had a hard struggle for existence both with wolf and man, and is now very rare and local. It is gratifying to know that it is being well looked after, and that the first observation of its unexpected horn-shedding was made by Abraham Bartlett at the Zoo in London.

As, however, the prong-buck's horns are only about a foot long as a rule, their shedding is a small matter compared with the same process in the deer, whose armament may measure several feet and weigh several stone. In fact, in the deer we come to the zenith of efficiency and of beauty in animal weapons —and of paradox, for a deer's " horns " have no horn in them at all. They are pure bare bone, though, when soft and growing, covered with the plushy-haired skin known as " the velvet." At this time they are sensitive, and this must help to teach the stag how to keep his head clear of the boughs, for one of the deer paradoxes is that these creatures with their branching crowns do not inhabit open country as a rule, but keep to the woods, while the hollow-horned ruminants with their simple horns are found away from them more often than not.

Deer appeared earlier in evolution than these rivals, and in some ways still are more primitive—there are,

W. S. Berridge

BUFFALO OF NEPAL AND WIDE-SPREADING HORNS OF AN OX OF SPAIN

Curved like scimitars, the horns of the wild Chittagong buffalo of Nepal (bottom) are terrible weapons indeed. The upper photograph shows a simple type of horn—smooth on the outside and built up round a core of bone, as all bovine horns are. The owners of these enormous horns are the Spanish cattle of which the above is a Zoo specimen. In the days when drinking cups were made from horns, such a pair as this would have been a real prize. As musical instruments, too, they suggest great possibilities.

W. S. Berridge

WILD GOATS: NUBIAN IBEX AND HIMALAYAN MARKHOR

Some of the ridges in the horns of the Nubian ibex (bottom) are so pronounced that these weapons seem almost to have a saw edge. This wild goat is famous for the backward bend of its horns. The markhor (top) is another species of wild goat and lives in the Himalayas. It is about the largest goat in the wild and possesses magnificent horns, each one formed in a bold spiral. Five feet is the measure of some of these spirals, and besides this the male has his head further enhanced by a handsome beard.

James

HORNS OF DEER, WILD GOAT AND ANTELOPE

Branching like a candelabrum the spiked antlers of the wapiti (bottom left), the American and Asiatic representative of the red deer family, are as dangerous as beautiful to see. The upper branches are probably used as guards when fighting, the deadly blows being given by the dagger-like tines on the forehead. The other photographs show (bottom right), the Grecian ibex, a wild goat which has very massive horns for its size, and two members of the antelope family, the African eland (top left) and the Arabian gazelle.

STRANGE HORNED SCREAMER OF THE SOUTH AMERICAN LAGOONS

There are several strange things about the bird called the horned screamer which is found only in South America. In the first place, it possesses a curious growth of horny substance, about five inches long, upon its brow. It is the only one of the five species of screamer which possesses such a thing. The birds haunt lagoons in enormous flocks and are in the habit of screaming all together in a chorus. When a number of flocks do this in turn the neighbourhood is apt to become intolerable.

for instance, two deer, the musk and the Chinese water-deer, which have never developed horns at all; nor do the females normally develop them, except in the case of the reindeer, and not always in this, for in Kazan reindeer hinds are hornless. The ugliest and most unpractical antlers are borne by the reindeer-like milou or Père David's deer, and have no tines in front, but a huge one at the back.

There is remarkably little difference in appearance between the little duyker-bucks of Africa and the little deer of South America known as brockets; only the little single-spike weapons stay on in the duykers and are shed annually in the brockets. Starting from these, all stages in the evolution of the deer antler may be found still existing, and paralleling the stages through which the young stag goes; for species with well-branched horns always begin with a single spike and add branches each year till the full coronet of antlers is assumed.

Next to the brockets we get the Chilian huemul, which has two spikes, like the Asiatic muntjac, which also has two, but mounted on high, bony supports. This is really a more primitive type, for it has well-developed canine teeth, and trusts to them in fighting more than to its little antlers.

A further evolution gives us antlers of the type borne by the Indian sambhar and spotted deer, with a brow tine near the forehead, a long beam or main shaft, and a couple of spikes on the top or crown. This is probably all that is required for efficiency, the brow tines being the daggers by which deadly damage is done, and the tops or crown-points the fencing weapons which keep the enemy from pressing his attack home.

But evolution appears to have run riot in deer's antlers as it has in the comb of the domestic cock, and gone beyond what is practical. For instance, in the beautiful antlers of the red deer, and his great ally the wapiti of eastern Central Asia and of America, we get a " bay " tine, which is merely a duplication of the brow tine and situated just above it, and a " tray " tine which, placed as it is about the middle of the beam, is hardly likely to come into action.

As an example, however, of the terrible efficiency of these weapons there may be quoted the case of a foolhardy trespasser who was killed in Caton's deer park in the United States by a wapiti stag, in spite of the efforts made to save him by several men armed with pitchforks which the stag kept off with the ends of his antlers while he gored his victim with the brow tines.

Casualties among deer themselves when fighting are remarkably rare. A deer duel consists mostly in head-to-head wrestling, and the beaten stag generally suffers by not being quick enough in getting away, and so laying himself open to a flank attack.

An occasional accident, however, brings death to both combatants—the interlocking of their antlers, which means death by starvation. This is particularly likely to occur in the common white-tailed deer of North America, which, like most American deer, has no brow tine, and has a forward-curved beam well provided with spikes. But it may even happen with the comparatively simple palmate antlers of the moose, and a pair of moose antlers thus locked are among the treasures of the Royal College of Surgeons Museum in Lincoln's Inn Fields, and date back to the time of the great John Hunter.

The palmate or broad-ended type of antler has been evolved independently in three types of deer—in the reindeer and the fallow deer, as well as in the moose. In the first-named even one of the brow tines is palmate, and is known as the " snow-shovel," though said only to be used as such when the snow is crusted, the deer usually digging with his fore-feet, like the bison, sheep, and horse.

Palmate horns in the stag's earlier years are of the ordinary branched type, and it is to the palmate form that the finest known antlers belong—those of the giant fallow deer dug up below the Irish bogs. Though extinct, this noblest of the horned animals lasted down to Stone Age human times. Its weapons may span nearly four yards across, with palms as big as the seat of a chair, fringed with spikes each of which would make a presentable horn for a deer of average size.

It is possible, however, that the record for antler length is held by a modern deer, which justifies the poor angler's ridiculed plea that the biggest fish get away, in that he was never shot or even seen. This was a wapiti stag whose shed antlers were found by an exploring party in the United States in the nineteenth century. So huge were they that when set up on their tips so as to form an arch the tallest man in the party walked under this without touching.

It is a pity that, as the specimens evidently could not be carried away, more exact details were not recorded; but as the man could hardly have been less than five feet eight, and as several inches at least must be allowed for the curve of the antlers, they must have been over six feet each, and were quite possibly seven.

After all this, it seems strange that a stag can get on perfectly well in deer society without any horns at all; but deer seem to delight in paradoxes, and a " hummel " stag which never develops antlers may be herd master, overcoming his armed rivals by boxing with his fore-feet like a horse. Hinds also fight in this way, but such a means of defence seems not to have been successful in the long ages of evolution, for those families of hoofed animals which have never developed horns contain very few species even compared with the deer, which are far less numerous and far less varied in bodily form than the hollow-horned section. They have, however, held their own far better, in spite of the drain on the constitutions of the males by the yearly re-growth of the antlers, and in spite of not possessing the enduring speed and extraordinary springing power so common in antelopes.

Chapter LXXIV

The Mystery of the Cuckoo

By Oliver G. Pike

Author of "The Great Winding Road"

For nearly two thousand years the cuckoo has been a bird of mystery. The result has been that an enormous amount of controversy has raged around this interesting bird. At the present day, students of the cuckoo are divided in their opinions; there are those who say that when she deposits her egg in the nest of the fosterer, she does it with her beak, having previously laid the egg on the ground some distance from the nest. Others say that sometimes she lays her egg into the nest, and at other times carries it there in her beak. A very small minority, who need not be taken seriously, insist that no cuckoo ever has laid her egg direct into the nest.

From the very earliest times the generally accepted theory has been the first one mentioned, but this was completely shattered when Mr. Edgar Chance, after several years of intensive field-work, showed that several cuckoos which he had under observation laid their eggs direct into the nest in the normal manner like any other bird.

Notwithstanding his very conclusive discoveries, he met with much opposition from those who disliked the old theory being upset. Some asked how the cuckoo laid her egg into a fragile nest like that of the reed-warbler. Others mentioned instances of cuckoos' eggs having been found in the homes of the wren and chiffchaff—two small nests that are domed over with a small entrance at the side, far too small for the cuckoo to enter. A few of the more violent critics said that it would be necessary to photograph every female cuckoo in England if the new theory was to be proved, for some might lay into the nest, while others would carry their eggs to deposit them with their beaks.

Following the publication of Mr. Chance's book "The Cuckoo's Secret," I worked in conjunction with Mr. Chance, and together we produced a kinema film which proved beyond every shadow of doubt that the cuckoos which we photographed laid their eggs into the nests, after first stealing one of the original eggs. A wonderful slow-motion film was obtained which showed every detail of the cuckoo arriving, stealing the egg, and laying her own egg in its place, afterwards flying off with the stolen egg which she retained in her beak while laying. A series of most conclusive pictures was obtained at five different nests.

Since then others have watched various cuckoos, and some wonderful observations have been obtained. It has been found that whenever the watcher had obtained a clear view of the whole performance, on each occasion the cuckoo has been seen to lay direct into the nest. Observers have seen her lay in the nests of the following half-dozen species of birds: tree pipit, meadow pipit, rock pipit, yellow-hammer, pied wagtail, and reed warbler.

I have seen a cuckoo flying round a hedge-sparrow's nest in which there were two eggs of the owner bird. I had a clear view of the cuckoo when she went to the nest, and there was definitely no egg in her beak. Less than ten seconds later the nest contained one hedge-sparrow's egg and one deposited there by the cuckoo. There was no doubt in my mind that she laid it. This rather imposing list contains most of the chief birds that the cuckoo uses as a fosterer, and it goes far to prove Mr. Chance's and my belief, that all cuckoos lay into the nest and do not deposit the egg with the beak.

Some very careful naturalists have made observations where eggs of the cuckoo have been found in small domed nests. I have watched the cuckoo lay many times; I have been very near the bird on each occasion, and I have come to the conclusion that she would have very little difficulty in laying into these nests by holding on to the sides with her claws, then laying the egg into the entrance. It has been noticed that when eggs are found in such nests, the front or sides of each has been disturbed where the cuckoo evidently clung to the exterior. At some of these nests the cuckoo's egg has been found just outside, which goes to prove that when she attempted to lay the egg through the entrance she failed to do so and it rolled down and remained outside. If she had been placing it there with her beak, and happened to drop it, she would not have been so foolish as to leave it lying outside, but would have picked it up and made another attempt.

On one occasion we surrounded a pipit's nest with thorns and twigs, leaving a little hole at the entrance through which the cuckoo could place her egg with her beak, but we made it so that the bird herself would have extreme difficulty to pass through the entrance. It was interesting to watch what happened when the cuckoo arrived. She first went to the entrance and found it was not possible to pass that way. In front was a flat space on which she could have easily laid her egg, to pick it up afterwards if she so desired. But the bird had no intention of doing this. She wished to deposit her egg in that nest, and the only manner in which she knew how to accomplish this was to lay it. She examined the gorse bush under which the pipit's nest was concealed, then she set about performing her allotted task. She settled on top of the bush and then, putting her head down, fought her way through the thorns. She lost several feathers as she went down, but reached the nest. I watched her first take one of the pipit's eggs, then she moved her body on to the nest, laid her egg and returned the way she came.

A. H. Willford

YOUNG CUCKOOS APPROACHING ADOLESCENCE

During the adolescent stage the cuckoo's plumage is brown on the upper parts with a white spot on the back of the neck and grey markings. In the upper photograph note the characteristic swelling under the beak which gives an incongruous effect as of a double chin to the young bird.

Another discovery goes a very long way to prove that this method is the one always used by the cuckoo. It was found that before laying her egg she sat for long periods on a branch of a tree or some other convenient perch. Here she remained almost motionless for anything from one hour to six hours, usually the longer period, and this corresponds to the time that other birds sit on their nests before actually laying. Directly she is ready to lay, she leaves her perch and *glides*, not flies, in a slow and careful manner to the nest. When she reaches it no time is lost. First she takes out one of the original eggs, quickly moves her body on to the nest and lays her own egg in its place, the whole performance seldom lasting more than ten seconds. She then flies off with the stolen egg, settles on a convenient perch, throws her head back and devours it. She swallows it rather ravenously, for if she has spent some hours on her perch before laying this is the first meal she has tasted for a long time.

It has been found that the cuckoo lays her egg every alternate day during the laying period. In the ordinary way she would lay about five eggs, and it is a great mistake to imagine that when she has performed this duty she forgets all about them. If her eggs are taken from the nest or destroyed in any way, she will do what other birds do under the circumstances, that is, lay again, and by taking her eggs directly after they were laid, one cuckoo was induced to lay twenty-five eggs in one season. She is not satisfied until she finds that her scattered brood is well on its way to being reared by the various fosterers. When success is assured, she and her mate leave the country.

THE story of the young cuckoo is one of the most wonderful in the annals of natural history. The period of incubation of the cuckoo's egg is thirteen days. When the youngster first hatches out it is blind, and, lying at the bottom of the nest, it appears as a little lump of black flesh. On the first day it takes no notice of its nest companions or other eggs in the nest, but on the evening of the second, or the morning of the third day, this helpless-looking infant becomes imbued with marvellous strength and instinct. The remaining eggs, or perhaps young in the nest, represent future rivals for food, and this the cuckoo cannot tolerate. So the youngster empties the nest of all other eggs or young.

If there should be eggs there, his task is fairly easy, for he gets one upon his back, stands up, and rolls it over the side. With young, his task is more difficult, but he quickly accomplishes it. I have often tested the strength of young cuckoos with birds very much larger and stronger than themselves. At one nest of the reed-warbler that I desired to film, I found that the young cuckoo had forestalled me, and the three young warblers were lying dead under the nest. I found a young sedge-warbler in a neighbouring nest, which had feathers on its body, was twice the size of the cuckoo, and was able to see what was taking place. The cuckoo was still blind, but directly I placed the warbler in the nest a most wonderful struggle began. As I watched I could see that it could have but one ending. The cuckoo showed the most amazing powers of strength. It worked down in the nest until its companion was on its back, then, gripping the sides of the nest with its feet, it used the wonderful muscles on its legs and slowly but surely raised its burden. When it appeared to have reached the extent of its stretching powers, it opened the small, fleshy arms that would some day be two fine wings, and began to work these up and down. It also jerked its body upwards with violent movements, with the result that the warbler was flung ignominiously over the side of the nest. This amazing youngster now tumbled to the bottom of its home, worked round to see that there were no more rivals present, and then settled down to contentment and rest.

M. H. Crawford

ADULT CUCKOO ON THE WING

Cuckoos are very fond of the various kinds of hairy caterpillars and are a definite check on the very destructive caterpillar of the garden tiger moth. This caterpillar is familiarly known as the "woolly bear," and will devour almost any kind of vegetation and, in some summers, becomes quite a local plague. The cuckoo's services as an antidote are then very welcome.

The parents of insect-eating birds bring food to the nest at very short intervals throughout the summer day ; often less than a minute passes between the visits. The young cuckoo, therefore, obtains an enormous quantity of food, and this is necessary, for it is a quick-growing bird. Ten days later it is clothed in fine feathers and is considerably larger than the nest on which it endeavours to make itself comfortable.

A few days later it opens its long wings and commences to fly, and as it travels from perch to perch the foster parents follow it. The baby will usually sit in fairly prominent positions, and occasionally utters a curious, high-pitched note. This seems to have a wonderful effect. Other birds in the district may be taking large beakfuls of food to their own young, but if they see the young cuckoo, or hear that peculiar cry, they pause in their flight, and give the food that they are carrying to the young cuckoo, and they will even return with further supplies. There was a record a few years ago of a cuckoo that was reared by a pair of hedge-sparrows being fed by five different species of birds, and I know of another instance in which no fewer than fourteen pipits were attending to one young cuckoo!

Cuckoo approaches nest

It seizes the egg inside

Holds the stolen egg in its beak

Settles on the nest

Lays its egg

Flies away with stolen egg

THE KINEMATOGRAPH CAMERA SOLVES THE ANCIENT CUCKOO MYSTERY

Until the kinematograph camera was invented and used in the naturalist's service the cuckoo mystery remained insoluble. But, with this new implement, the investigators were, after the expenditure of much patience, rewarded with conclusive evidence from the films they obtained. The bird waits near some selected nest until the time for laying her egg and then glides to the nest, removes an egg with her beak, lays her own very small egg and flies off with the stolen one which she devours. This story is fully told here in photographs by Oliver G. Pike.

CUCKOO CAUGHT IN THE ACT OF LAYING HER EGG

Having selected a meadow pipit's nest as a suitable place for the incubation of her egg this female cuckoo has flown to the nest and seized one of the pipit's eggs which we see here held in her beak. She is just in the act of laying her own egg, which is only about the size of a lark's, into the nest in place of the one she has just stolen. This egg she will soon swallow, for she has not tasted food for some hours while waiting on some nearby perch for her own egg to be ready for laying. This photograph is from a kinema film by Oliver G. Pike and E. P. Chance.

Oliver Pike

HARD-WORKED SEDGE-WARBLER ATTENDING TO HER GARGANTUAN CHANGELING

It is at once a marvellous and ridiculous sight when a young cuckoo, three-parts grown, is being fed by one of its foster-parents who may be only about one-third of its size. In the left hand photograph the young cuckoo, which is in the stage when it flies short distances from one perch to another and is followed about by its foster-parents, is sitting on a branch and looking up expectantly to the warbler who is always a giver of good gifts. On the right the cuckoo is showing the commodious dimensions of its mouth while the warbler is standing, with one claw raised as though in uncertainty, gazing at its insatiable great charge and, perhaps, wondering how this all happened.

A. H. Willford

LUCKLESS MEADOW PIPITS FEEDING THEIR HUGE FOSTER-CHILDREN

On the right is a photograph of a common sight where young cuckoos are concerned, the foster parent standing on the young bird for the sake of convenient feeding. The other illustration is remarkable for showing two cuckoos at once on the same stump. There is a record of no fewer than fourteen pipits attending to one cuckoo, and this despite the fact that they all had healthy, hungry young ones of their own squalling for nourishment at home. This remarkable bird seems to have the secret of charming other birds from their proper duties.

A. H Willford

Oliver Pike

SEDGE-WARBLER FOSTER PARENT AND A MEADOW PIPIT'S CHANGELING

Cuckoos seem to prefer certain birds as nurses for the chicks they do not trouble to rear themselves. Sedge-warblers, pied wagtails, meadow pipits, and hedge sparrows—all small birds—are the most popular for the purpose, and as the cuckoo lays a comparatively small egg this can be accommodated in the nests of these birds. The lower photograph shows a sedge-warbler with a beak-full of insects, feeding the cuckoo which seems almost large enough to be able to swallow its nurse—insects and all. Above is a young cuckoo in a meadow pipit's nest.

Beginning of the struggle

The cuckoo gets underneath

The cuckoo spreads its wings

The rival pushed over the side

Going

Gone

THE CUCKOO IN THE NEST AND HOW IT BEHAVES THERE

So soon as ever the cuckoo is hatched out, it proceeds at once to dispose of any rivals. It needs all the available food supply and takes at least two birds, working whole-time, to feed it. If it finds it has hatched first, it heaves the other eggs out of the nest. If there are other fledgelings it adopts the method illustrated by the kinematograph illustrations above. The cuckoo gets underneath its rival and uses its strong legs to press itself upwards, thus lifting the other bird. Then, by agitating its wings, it hustles its rival over the edge. Photos by Oliver G. Pike.

A. H. Willford

CUCKOO BEING FED BY MEADOW PIPIT

The cuckoo is specially equipped for its nefarious system of up-bringing. It soon grows its feathers and starts to fly. The parents follow it about and bring food. Not content with this the cuckoo utters, from time to time, a peculiar cry which seems irresistible to other birds, for they come and bring it the food they were taking to their own chicks.

But at length the time comes when the foster parents become tired of following and feeding their great infant, and they leave it to get on as best it can. But now it is confronted with a difficulty. Most young birds, when they leave their nests follow their parents about, and learn how to collect their own special food, but the young cuckoo has been content to have all the food brought to it, and when it has to obtain its own living it discovers that insects, unless you know where to look for them, are difficult to find. It becomes very hungry, and eventually changes its diet and becomes a vegetarian.

The crops of twelve young cuckoos that had been deserted by their foster parents were examined, and no trace of insect food was found. Eventually these youngsters, that are so cleverly able to adapt themselves to their surroundings, find where the insects are hidden, and go back to their proper food. Five, eight, or even ten weeks after the real parents of the cuckoo leave the country these youngsters launch out on a long southern journey, eventually to reach Africa, where they spend the chief part of the year.

There is still much to be discovered about this interesting bird. But a word is necessary as regards those observers who say they have seen cuckoos carrying their eggs in their beaks. The cuckoo is a confirmed egg thief, and on her non-laying days spends quite a lot of time in searching for and raiding nests. I have watched her take and devour eggs from the nests of the yellow-hammer and skylark, and a casual observer might easily think that she is carrying her own egg, while as a matter of fact she is simply carrying the food she has discovered to a comfortable and convenient perch.

Many observers have seen a cuckoo standing at a nest with an egg in her beak, and with the old theory firmly implanted in their minds, they have thought it was her own egg that she was about to place in the nest. If these instances could have been properly investigated, I firmly believe that in every instance it would have been found that the egg she was holding was the one she had stolen from the nest previous to laying her own in its place. No one, as yet, has been able to bring forward satisfactory photographic proof, that is, with Kinema films, that a cuckoo places her egg in the nest with her beak.

When we sum up all the evidence we cannot help asking whether the naturalists of the past have not made a great mistake, and been rather prone to follow in the footsteps of their predecessors, accepting their statements without taking the trouble to investigate for themselves. In the light of modern discovery, observers of all kinds should start afresh, obliterating from their minds all previous theories. It is only by doing this that we shall be able to steal the cuckoo's secret from her, and prove a much debated problem.

It was not until the invention and perfection of the Kinematograph camera that so vexed a question as this could be really satisfactorily solved. One by one, the wise saws of old give way to modern instances—obtained with the help of ever increasing science

Chapter LXXV

Living Rainbows

W. P. Pycraft

Author of "The Seashore"

WHY it is that some animals have developed coloration which resembles the iridescence of certain metals no man can say. And the matter is rendered the more remarkable from the fact that this gorgeousness is not due to pigments answering to the hues displayed, but to a modification of the surface-structure of the feather, or hair, or scale, as the case may be.

How this metallic appearance is produced in a feather may well serve to illustrate the similar cases in other creatures. In birds then, where these shimmering hues are present the "cilia," or thread-like investment of the surface, are wanting; and the metallic areas are formed of rows of compartments covered with a transparent horny layer, often overlapping like tiles: and this transparent layer is backed by a black or brownish-black pigment. In some cases the outer case is smooth, but it may be polished or engraved with fine, wavy, longitudinal ridges; or small dot-like irregularities may take the place of lines. But in all cases the outer transparent coat acts just like a number of prisms.

All metallic feathers appear black when the surface is parallel to the rays of light and at the same level as the eye and the light. As one changes the incidence of the light, as in the king bird of paradise for example, the feather passes from black to light coppery red, then rich green: while in the beautiful *Pharomacrus moccino* —one of the trogons— it changes from black to greenish-bronze, through golden-green to green, from green to indigo, then to violet. In one of the humming birds, *Oreotrochilus chimborazo*, the whole solar spectrum is displayed—violet and red on the head, followed by orange and green on the back—blue, violet, and lastly, purple on the long tail feathers.

Most of us, I think, when we see the limelight thrown on a performer on the stage, are moved rather with the effect produced: and take no thought for the mechanism which produces it; and so here, it is the sheer beauty of the vivid and ever-changing hues which fascinates us, and not the wonderful underlying mechanism.

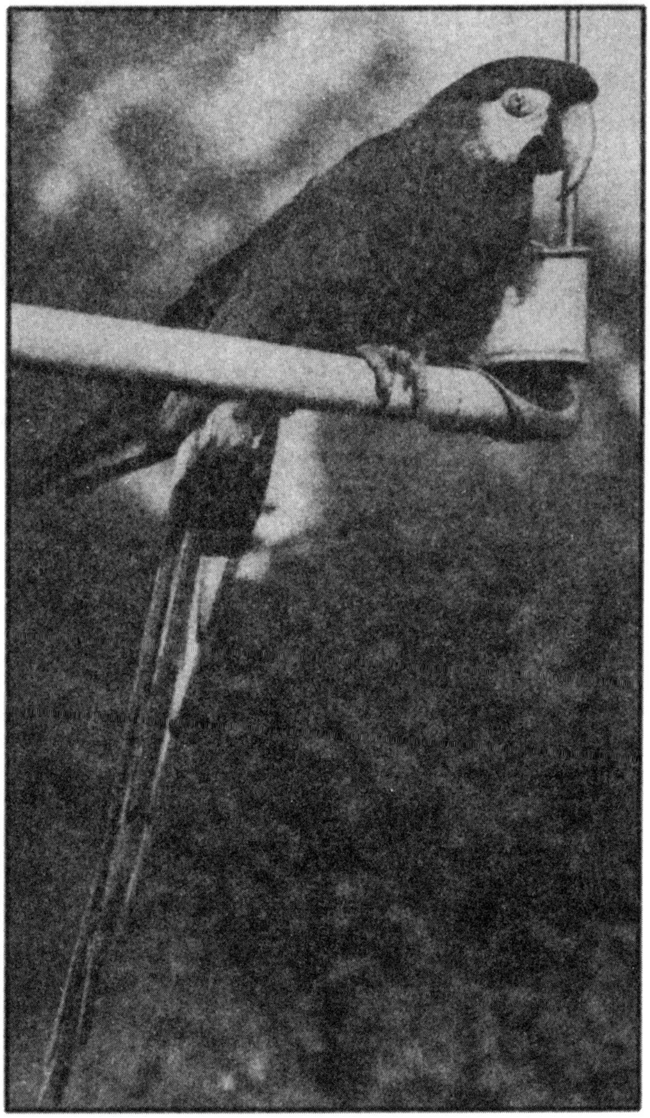

F. W. Bond

MACAW AND ITS RADIANT FEATHERS

This is a red and yellow macaw which is native to the Amazon forests and the northern parts of South America. The prevailing colour is a dark, rich red and its brilliant plumage makes it a favourite in aviaries. It will live ten years or more in captivity.

But these spectacular effects gain an added interest when an endeavour is made to discover what lies behind them, so to speak. Now in birds we find these effects always associated with the males, which seem not only conscious of their splendours, but convinced that, properly displayed, they will prove a sure and certain means of securing the favour of their mates. This may, indeed, be an inference well founded. It would not, however, be safe to conclude that all animals thus resplendent are animated with a like desire, for there is no evidence that the exquisite hues of certain butterflies, to be mentioned presently, are in the slightest degree appreciated by the wearers, or by the females. But we must take our facts as we find them, and not strive to make what is evidently true of one group of animals apply equally to some other

W. S. Berridge

BRILLIANT HEAD OF THE BLUE AND YELLOW MACAW

This bird was noticed by sixteenth-century explorers of South America, and its bright uniform of orange-yellow and green together with its conversational powers make it one of the most popular of the macaws to-day. The forehead is an olive-green and the back of the head green-blue, while the cheeks and throat are marked with a black stripe. The sides of the head are yellow. The blue and yellow macaw is found between Bolivia and Honduras and is essentially a bird of the tropics. It is very quick, as a rule, to pick up words and phrases.

group which we know as holding a widely different position in the scale of life.

The birds, as I have said, seem conscious of their beauty. Look at the "peacock in his pride." Having set his gorgeous train in order, he approaches his apathetic, and seemingly unappreciative, mate, backwards; so that she sees nothing but a great, grey screen. As soon, however, as he deems himself sufficiently near, he swirls round and faces her, then sets all the "train" feathers vibrating so that they give out a sound like the pattering of rain on leaves, and caps the performance with a wild scream! If we need evidence of "consciousness" of beauty, and a deliberate desire to exploit it, we seem to have it here.

Turn again to the Amherst and golden pheasants. I need not attempt to describe their splendours in detail; but will instead draw attention to that wonderful "frill" of exquisitely coloured feathers which surrounds the neck. When they are "courting" this frill is made to play a great part. The bird is an accomplished artist. He presents one side of himself to his mate, and gives a twist to that frill so as to bring most of it over to the side facing her, spreading out the long and resplendent tail at the same time. But note it is not spread like that of the peacock, but kept close to the ground, and spread sideways, while the wing on that side is half opened. Here, again, there seems to be a deliberate, conscious effort to make the most of his charms.

I know of no records of the "display" of the impeyan pheasant of Kashmir, which furnishes one of the most striking examples of metallic plumage among birds, for every feather of its body looks as though it were made of iridescent metal.

When we turn to the birds of paradise we are faced with such a galaxy of splendour, and such amazing methods of displaying these splendours, that the task of describing them almost induces a state of mental paralysis! Yet I would attempt the feat were it not for the fact that I want rather to draw attention to broad facts than to dwell upon selected cases.

Why is it that among some groups of birds, like the birds of paradise, the sun-birds, and the humming

MANDARIN DRAKE, A LIVING RAINBOW OF PLUMAGE, AND HIS MATE

In birds it is usually the male which displays the more brilliant livery, and this is generally heightened in beauty during the mating season. The mandarin duck is found in Eastern Asia and the male is particularly remarkable for its wonderfully tinted plumes. One of the scapular or shoulder feathers is upturned to form a fan upon the back, and this fan is a bright brown and banded as though with imperial purple.

PEACOCK'S FEATHER WHOSE SPLENDOUR IS NOT DUE TO PIGMENTS

The remarkable iridescent hues seen in a peacock's feather are not due, as might be supposed, to the presence of any pigments answering to the tints displayed, but to a modification of structure in the surface of the feather. That there is no pigment present can easily be proved by holding the feather parallel to the rays of light falling on it, when it will appear black. The parts of the feather which display the metallic colouring are formed of rows of minute compartments covered by a transparent layer and this layer acts like a number of prisms, thus producing the rainbow effects.

birds, almost every species, in the males at any rate, is thus arrayed in "purple and fine linen." I say almost every species advisedly, for, as is the case of the humming birds, there are some which are clad in "hodden grey," both male and female: as are the females of many of the most empurpled species. It is to these simple types that we must turn in searching for a clue to the evolution of the splendours. But that is another story.

Before leaving the birds it will be helpful to turn for a moment to species like the kingfishers, which present a striking array of beautiful forms, and some are superbly coloured. The common British kingfisher is an example, and here both male, female, and young wear a resplendent dress.

The coloration changes with the incidence of the light, passing from green to blue as the body turns first with the beak then with the tail towards the eye. But there is no "metallic" effect. Such colours belong to the group known as "objective" or "structural colours"; that is to say they are produced by the combination of a certain pigment with a superimposed, and very specially modified, but quite colourless surface.

Finally, be it noted, the variety of colours which may be borne on one and the same bird is often striking. Take the beautiful grass-finch, for example, which displays blue, yellow, green and red. No single flowering plant displays such a wealth of colour as this.

The gorgeous metallic-blue butterfly of the South American genus Morpho, with wings six inches across, one of the most beautiful of all insects, owes its superbly shimmering magnificence to the fact that every scale is engraved with fine, closely set, parallel lines or ridges. Under the microscope these scales, properly illuminated, are exquisitely beautiful objects. And the same is true of parts of the elytra, or wing cases of many beetles, especially, for example, of the wonderfully coloured diamond beetle of South America (*Curculio imperialis*). When seen under a low power this shows like clusters of jewels flashing against a dark velvet ground. The beautiful golden iridescence of the beetle Cassida, and its allies, is due to a film of

Peter Webster

JAY WHICH SOMETIMES SUFFERS FOR ITS BRIGHT COLOURS

The delicate art of tying artificial flies for fishing requires numerous kinds of feathers and the brilliant metallic blue feathers in the wings of the jay cause it to be hunted and slain to provide the angler with lures sufficiently bright to imitate the sheen upon certain insects that fish like. But it is not only the jay's wings which are brightly coloured. The crest feathers are white and tipped with black, and the nape of the neck and the ear coverts a rich, ruddy brown. A fine sight this bird makes, flashing in the patches of sunlight as it flies between the branches.

EXPANDING NECK FEATHERS OF THE AMHERST PHEASANT

An exquisitely tinted frill of feathers clothes the neck of the male Amherst pheasant. When the mating season comes round this frill is brought into play for the benefit of the female. The bird stands sideways to his intended mate and twists the frill round and up so that the head is almost hidden by a fan of gorgeous plumes. Through this the bird watches the effect of all this beauty on the female who, as a matter of fact, usually seems quite unmoved by even such magnificence.

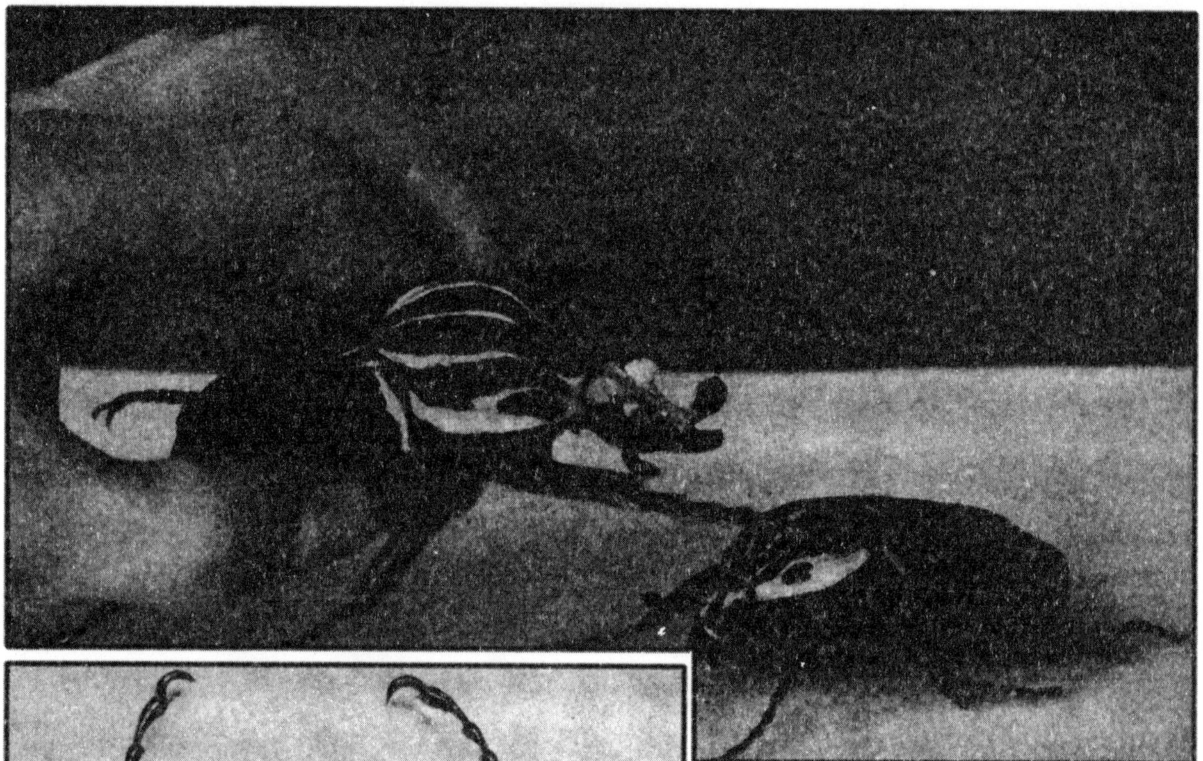

H. Bastin

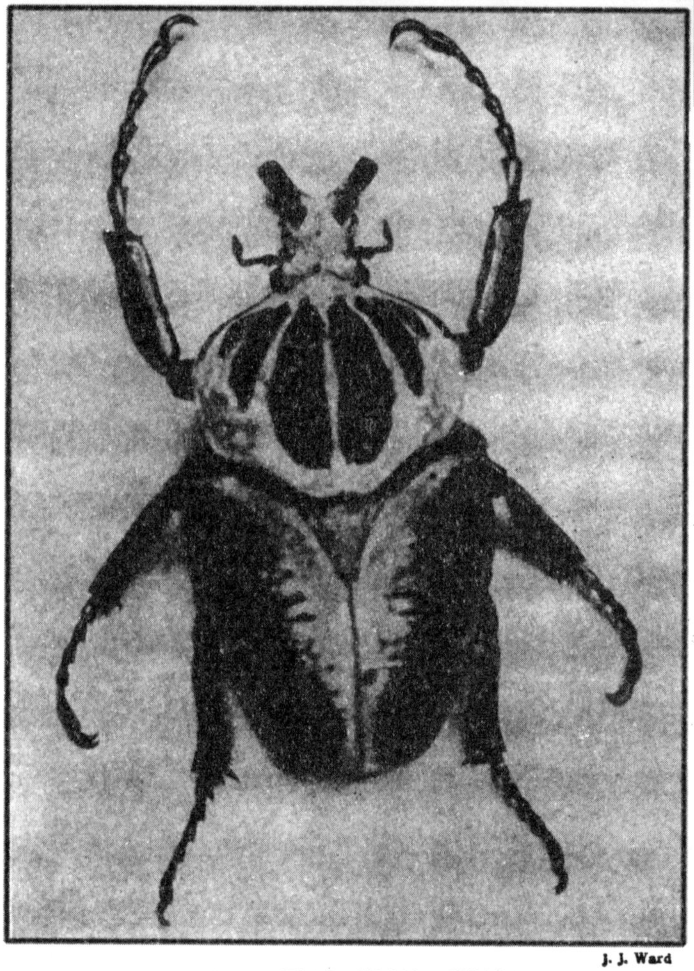

J. J. Ward

GIANT BEETLES FROM AFRICA

Tropical Africa colours many of her insects as gorgeously as her birds, and here we have some beetles, that in the lower photograph from Central and those in the upper illustration—goliath beetles—from West Africa. Some of these creatures measure about six inches from leg-tip to leg-tip.

moisture beneath the surface of the elytra; and hence it is that dried specimens which have lost their colour can be restored by soaking in water, provided that the drying has not been too prolonged.

So far this discussion has been confined to a fixed and definite coloration, which can in no way be changed by the wearer save in so far as, in moving and turning about, it takes up a new position in regard to the incidence of light. For the source of this coloration is embedded in hardened tissues like feathers; or scales, as in the case of butterflies. When, however, the pigments are, instead, deposited in the naked skin, matters assume a very different aspect, for they are now more or less directly under the control of the nervous and muscular systems.

The foundation of this coloration is made up of various kinds of "chromatophores," that is to say, of little "blobs" of colour contained within a delicate sheath or pellicle, having a radiate form, like, say, a starfish, their arms interlocking much as one interlocks one's fingers in clasping the hands. To the end of each finger muscular fibres are attached, and when these shorten, under the appropriate stimulus of light, the sheath flattens and spreads out the contained pigment. The shortening process is carried out by the stimulus of the nerves, and as these are

W. S. Berridge

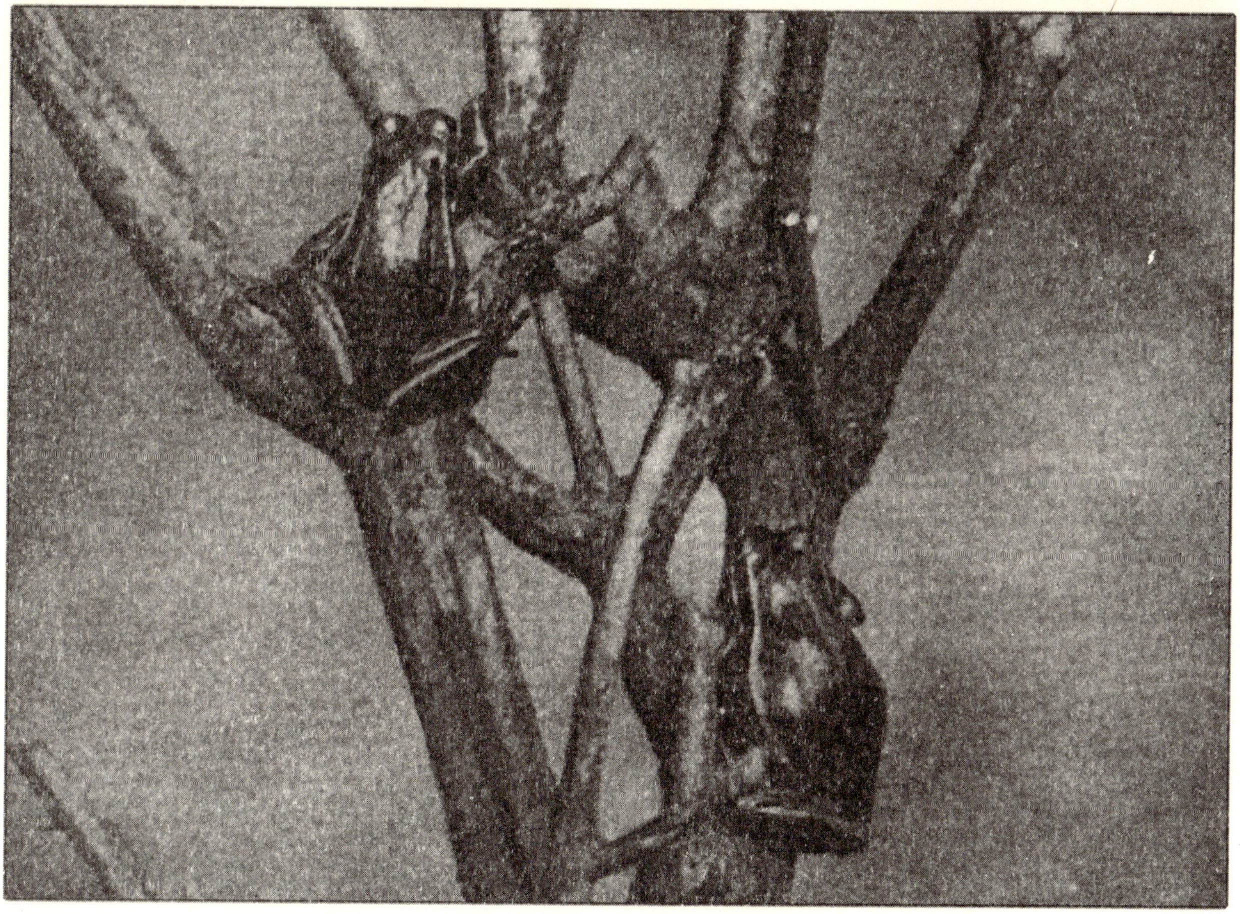

TREE FROG, CARDINAL AND GORGEOUS IMPEYAN PHEASANT

Frogs seem to favour bright colours and about the most brilliant is the golden tree frog (bottom). South America produces a brilliantly coloured bird, the red-crested cardinal (top left) whose name suggests the splendour of its appearance. But the most brilliant of all the world's birds is the impeyan pheasant or monal (top right), which is found in the high forests of the Himalayas and other big Asiatic ranges. The male's crest is green, shot with purple and blue, and the rest of this creature of colour is a blend of copper-red, golden green, crimson and white.

Bett

D. Seth Smith

BIRD OF PARADISE, MANDARIN DUCK AND HUMMING BIRDS

With the birds of paradise we are faced with the most amazing splendour of wonderful colour, the whole effect being enhanced by the strongly metallic sheen on the feathers. Here (bottom left) is a greater bird of paradise engaged in " displaying " for its mate. The lower right-hand photograph shows one of the handsome mandarin ducks of Eastern Asia, a fine specimen of which is seen in the colour plate in this chapter. The upper photographs are of those fluttering jewels, the humming birds of the tropics.

F. W. Bond

PALE-HEADED CAIQUE AND ALEXANDRINE PARRAKEET

The pale-headed caique (bottom) is a South American parrot of most brilliant feather, and belongs to a genus Caica or Deroptyrus. The colours are a gorgeous arrangement of green, white and orange. The upper photograph is of an Alexandrine parrakeet, whose colour scheme is made up of green and several different shades of very dark, almost black, feathers. This is native to Asia and parts of Africa, and the green plumage is almost invisible against the foliage of trees. But when against some contrasting background the brilliance of colour is very striking.

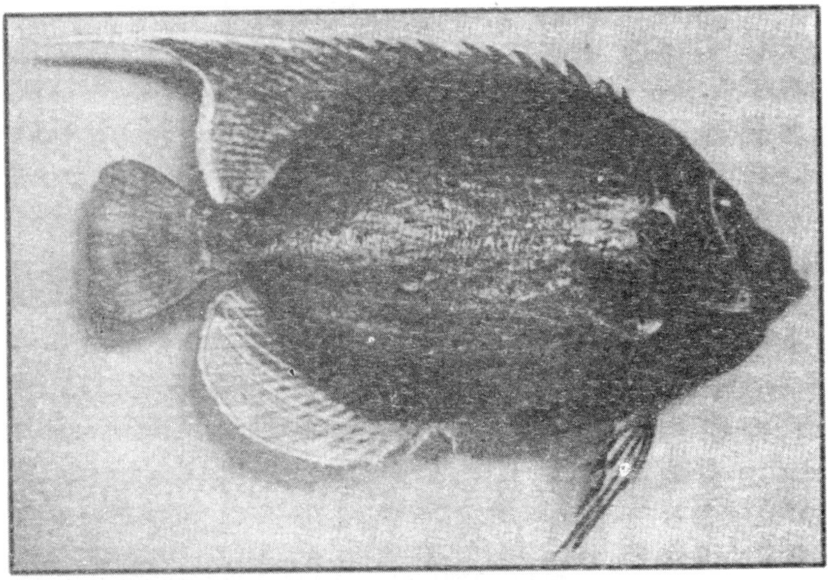

W. S. Berridge

BRILLIANT FISH OF WARM SEAS

Fish of the warmer seas tend to outdo all terrestrial creatures in the matter of colouring. This is not surprising when we consider the less transparent medium in which they live. Below is a long-spined chaetodon and above an "Emperor of Japan" fish.

"tuned" to different intensities of light—each pigment responding to a different stimulus, sometimes all the dark, and sometimes all the light pigment may be dominant.

Besides the black and coloured "chromatophores" just referred to the skin of the Amphibia—the frogs, toads and newts—contains yet a third element. This is represented by certain bodies known as "iridocytes" composed of a reflecting substance.

The nature of the changes effected can best be grasped by following what happens to the animals coloured after this fashion. The late Dr. Hans Gadow tells how he received a number of green tree-frogs (*Hyla arborea*) packed in moss, from a friend. When unpacked they were of a dull greenish-grey. They were placed in a cage with freshly-cut branches of a lime tree; next morning the leaves had withered, and the frogs were nowhere to be seen! But after a little hunting they were found sitting on the dark brown branches. They had assumed a mottled brown coloration so like that of the boughs as to render them at first invisible! Later, placed with growing plants, they climbed up into the branches and matched the green leaves.

Now there is no green pigment in frogs. This colour is an "interference" colour, formed by the aid of the upper "iridocytes" just referred to. These have a yellow upper-surface and a white under-surface. The light passing through the yellow layer is reflected by the white, thus giving rise to the green appearance.

Our first conception of these comparatively rapid changes of colour was derived from the chameleon, which has long enjoyed fame on this account. At night its prevailing hue is creamy yellow, by day a greyish-green with darker specks and patches. But under the influence of excitement these tints are intensified, the pale-brown patches deepen to a full maroon, while the green is dotted with golden yellow.

Nowhere is this type of coloration seen to better advantage than in the fishes. A mackerel, newly hauled out of the sea into a boat, presents a play of iridescent colour exquisite beyond words.

THE "flat-fishes"—soles, plaice, turbot, and so on—have formed the subject of some interesting experiments made by Mr. J. T. Cunningham. Normally, the upper surface of these fishes is some shade of brown, to match the sandy bottom where they rest, partly covered with the sand which they throw up by the movements of the fins, while the under-surface is white. By placing them in a tank with a glass bottom, and lighting this from beneath, Mr. Cunningham caused the white underside to become coloured like the upper. Some American naturalists went further than this in their experiments. A brill was placed in a tank floored with a pattern in black and white squares, like a draughtboard. This was all the fish had to rest on. As a consequence it speedily did its best to match this most difficult background, by developing patches of white all over the upper surface.

Chapter LXXVI

The Hidden World in the Soil

By Dr. B. A. Keen

Assistant Director of the Rothamsted Experimental Station

If a man were asked to think of the greatest contrast to our busy civilization he would almost certainly picture the countryside on a still summer afternoon, when all Nature appeared to be gently dozing in the warm sunlight. But he would be quite wrong, for this apparently immobile mask hides an amazingly active world—the hidden world in the soil.

Men of science are gradually unravelling some of its complications, not only because of the fascination it has for the scientific mind, but also because of the possibilities of improving agricultural and gardening methods that may be opened up by their discoveries.

We now know that the soil is not inert and lifeless. It is not just a mere collection of tiny broken fragments of rock (like the sand on the sea shore) but teems with myriad forms of life, vegetable and animal, in enormous numbers. We know also that without the activities of this population that inhabits the soil, the growth of crops, and indeed human life itself, would not be possible. Although we have as yet scarcely touched the fringe of the subject and its possibilities, enough has been found out to enable us to trace roughly the most important phases of the interlocked activities of the soil population.

This population includes many different forms of life. There are worms, insects of all kinds, and a bewildering variety of more lowly organized forms known as fungi, protozoa and bacteria. Some of these are helpful, some harmful, while others, as far as we can see, are neutral. Nothing like a complete census of the soil population has as yet been attempted, but even the incomplete figures are very large. Taking the larger forms of life, something like fourteen million separate individuals are found in the top nine inches of an acre of good soil, and most of these are concentrated in the top few inches.

The method used to obtain these figures is simple, although somewhat laborious. A block of soil of a known size is taken from the field, and placed on a coarse sieve, that stands on the top of a series of other sieves with finer meshes. The soil is broken up into smaller pieces and then a stream of water is directed on to it. All the fine soil particles are washed away, leaving on the sieves only the stones and coarse soil particles, together with insects and other members of the soil population. These can then be easily picked out, identified and counted. About half of them are insects, the remainder being worms, millipedes, centipedes, slugs, and so on.

But these numbers are completely eclipsed by those for the lower forms of life. The bacteria alone reach enormous figures. In a quantity of rich garden soil that could be easily held in an ordinary saltspoon there are no fewer than 5,000 millions of these organisms present, a number which is more than twice as great as the human population of the whole world. It seems almost incredible, for our first feeling is that there could not be room in the small quantity of soil for such an enormous number. But when the minute size of these organisms is taken into account the numbers are seen to be quite possible. The average diameter of a single organism is only about the ten-thousandth part of a centimetre, or the twenty-five-thousandth part of an inch. A simple calculation shows that the volume occupied by them is only one per cent. of the volume of the soil itself.

It is not easy, however, to visualise such small sizes and such excessive numbers as those just mentioned. We can get a better idea of the size of bacteria in relation to other members of the soil population if we imagine everything to be magnified to dimensions with which we are more familiar. If the bacteria were each as big as a pin's head (about $\frac{1}{16}$th of an inch in diameter) then on the same scale an earthworm only two inches long and $\frac{1}{12}$th of an inch thick would become equal to a Tube train, and he would travel through his burrow at about the same speed as the train does through its tunnel.

Although much more could be written about the soil population as a whole, the chief interest centres in what the main groups do and to what extent the growth of plants is dependent on, or hindered by, their activities.

Earthworms have been recognized as beneficial agents by all gardeners and farmers for a very long time. They

COCKCHAFER GRUB

For three years or longer the grub of the beetle called the cockchafer lives under the turf or the ploughed land, well hidden in the soil, feeding on roots and doing enormous damage to man's agriculture.

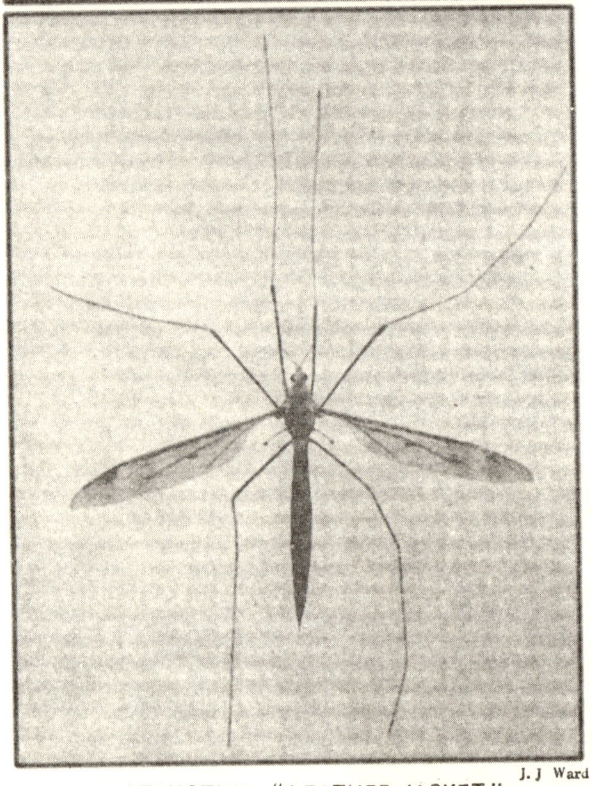

J. J Ward

DESTRUCTIVE "LEATHER-JACKET"

The crane-fly is a dainty creature to see, but its progeny spend their larval stage in the soil where they are known to farmers as leather-jackets. Below we see the fly itself, and above the repulsive grub which does so much damage to growing crops.

indirectly benefit the plant by improving the condition of the soil itself. Charles Darwin believed that there were about 50,000 worms in an acre of rich garden soil, but recent experiments show that the number is more nearly a million, or one for every six square inches of the soil surface. Their burrows provide an easy means of entry for rain into the soil; their habit of pulling leaves and other vegetable material into the burrows provides the soil with vegetable mould and thus improves its fertility.

The "worm casts" on the surface of the soil consist of numerous small particles of soil that have been passed through the body of the worm in the process of feeding. These casts gradually break down under the alternations of rain and dry weather into a crumbly layer so that the soil surface is continuously being renewed by particles brought up from below. Darwin concluded that a layer of soil about $\frac{1}{50}$th of an inch thick was brought up each year by this means, so that any object originally on the surface would in the course of time become covered with soil. The thick layer of soil over the Roman pavements in this country has largely accumulated in this way.

Although earthworms are welcomed in the flower border or the kitchen garden, their presence in the soil of tennis courts, bowling and golf greens, is regarded with mixed feelings. It is difficult to maintain a detached philosophic attitude when a truly

WORM HOLES IN SOIL

Here is a section of soil cut from a field. It shows the number of tunnels made by worms in even a small area and illustrates how helpful these passages are for distributing rain. The photograph is from Russell's "Lessons on Soil," Cambridge University Press.

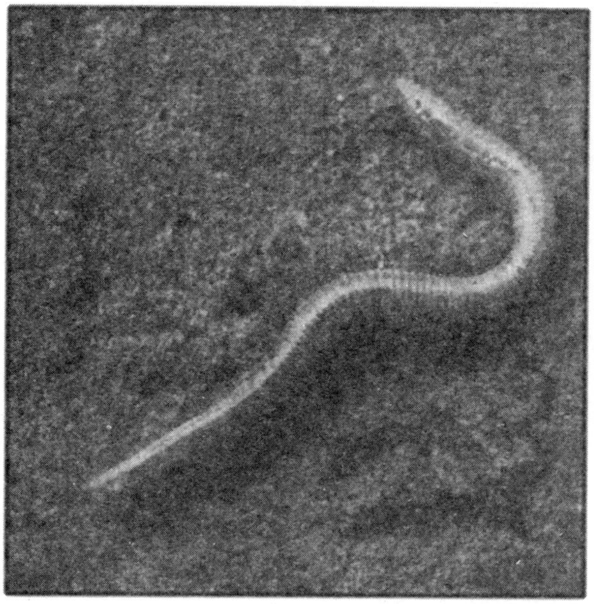

J. J. Ward

WONDERFUL EARTHWORM AND ITS LEAF-BURYING HABIT.

On the left we see an earthworm discovered by the camera in the open. The sun shining on its moist body makes it glitter like a piece of armoured hose. Worms continually aerate the soil by their burrows and renew the surface with their castings—the earth which has passed through their bodies. The right-hand photograph shows some of these castings and a dead oak leaf partially drawn into a worm's burrow. This habit of pulling dead leaves under ground for food provides leaf-mould to the soil.

struck ball is deflected by a worm cast. Certain chemical substances are known that are injurious to worms without seriously affecting the grass, and these are usually applied to the lawn in a water solution. Fortunately the copious fibrous roots of the grasses that ramify through the soil produce the same general action as the worm burrows and casts, so that the physical condition of the soil—and hence the growth of the grass itself—is not seriously affected by destroying the industrious worms.

THE next great group to consider is the insects. Some pass their whole life in the soil, while other species occupy it only during the pupal stage. All animals are dependent on plant life for their food; either they consume it directly, or, if they prey on other animals, indirectly. Usually there is a balance between the different groups, any tendency for one group to increase in numbers being checked by a corresponding increase in those groups that prey on the first group, until the equilibrium is re-established. Occasionally the check fails; conditions may become so especially favourable for the development of one form of insects that they increase greatly in numbers in spite of everything. If this particular species lives on vegetation it becomes a severe plant pest and may destroy large areas of crops, until it is brought under control, either by natural agencies or by the extensive use of suitable insecticides. The study of injurious soil insects and their habits is very important in order that their most vulnerable period may be known and chosen for attack.

One very unexpected result of the speeding-up of ocean transport, when fast steam vessels replaced the old windjammers, was the outbreak of new insect pests in some countries. The probable reason is that any such insects in the cargo are more able to survive the shorter voyage. They may in this manner obtain a footing in a country formerly free of them, in which their natural enemies may also be absent. In these conditions they will multiply unchecked, and soon become a serious menace. A most interesting recent example of this is the infestation by earwigs of parts of New Zealand, and the question at once arose, why should they multiply excessively there and not in England, when the general climatic and other conditions are so similar in the two countries? Investigation showed that in England earwigs are kept in check by a parasite that was absent in New Zealand. Numbers of this parasite were specially bred in England and shipped to New Zealand in the pupal stage to be liberated there and combat the pest.

AN earlier instance of a disturbance to the natural balance of insect and plant life was provided in Honolulu, where the extensive sugar-cane plantations were threatened with extinction by the great increase in numbers of certain leaf-hoppers, and stem-boring insects. All attempts to check these pests failed, until the natural parasites were especially introduced into the island, and in a very short time the trouble disappeared by the restoration of Nature's balance between the pests and the parasites.

On the whole, if we exclude the vital function of insects in pollination of plants, we may say that they are injurious rather than helpful to plant growth, so the gardener and farmer are more concerned to destroy than to preserve them. They provide in the soil a certain desirable stirring of the particles—comparable

J. J. Ward

BEETLES THAT ARE BUSY SCAVENGERS OF THE SOIL

Often found in gardens is the violet ground-beetle (left) which is to be distinguished by the violet-coloured edge of its wing cases which are otherwise black. This beetle should not be touched, as it is liable to eject, for its defence, a foul-smelling liquid. The sexton beetles (centre) get their name from their habit of burying any dead bird or mouse they find for the purpose of feeding on it and laying their eggs in the body. The right-hand illustration is of a dor beetle among the grass. This is another scavenger of the soil.

to the action of worms—and, as a kind of poetic justice, their dead bodies become available in due course as plant food.

This leads us logically to a consideration of how animal and vegetable material become available for the nutrition of plants. It is here that the minute types of the soil population—the bacteria, protozoa and fungi—play a fundamental part, for they are actively concerned in the process of converting old plant residues into simple substances that are soluble in water, and therefore in a state to be taken up by the roots of the growing plant

In the first place, however, we must get some idea of what these small organisms are, how they exist, and of the methods the scientist uses for studying them.

Bacteria really belong to the family of plants; they are not animals, but consist of single cells that have similar characteristics to the enormous number of cells of which the higher forms of plants are built up. When seen under a powerful microscope they look something like small transparent lumps of jelly, usually of a simple shape—round, oval, or squat rods. Some kinds have one or more waving, hair-like projections, by which they propel themselves through

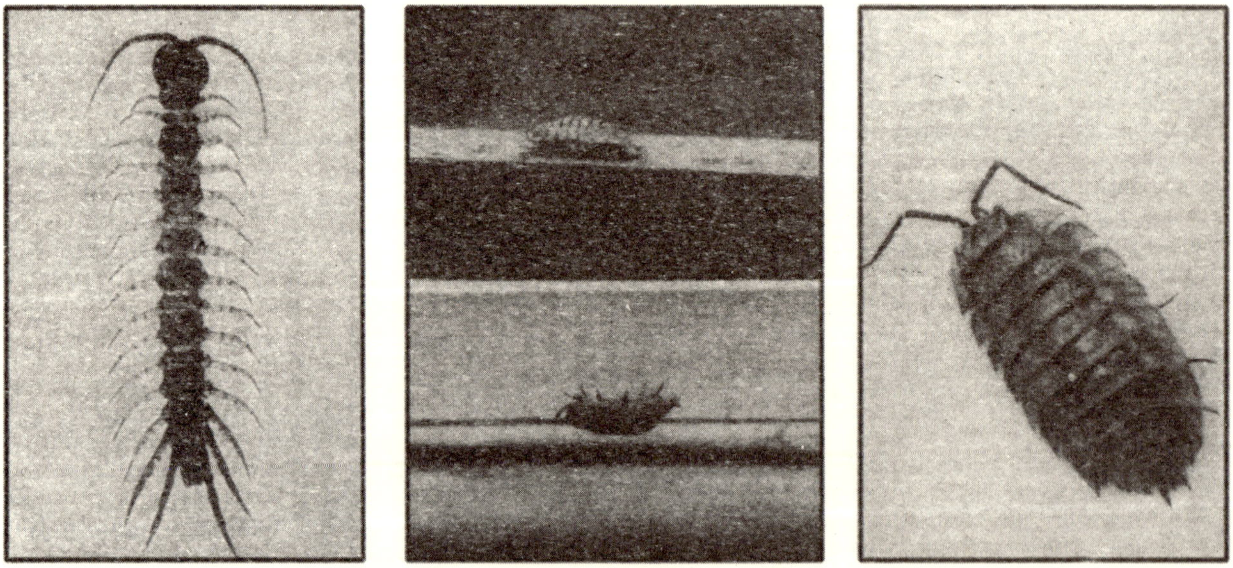

CENTIPEDE AND WOOD-LOUSE OF BRITISH SPECIES

Here (left) is a British centipede which is commonly found under stones. The number of legs in the centipede varies greatly, and a pair is attached to each segment of the body, the number of segments varying between fifteen and one hundred according to the species. The other photographs show a species of wood-louse (*Porcellio*) found in Britain, that on the right being magnified. This frequents damp places, but, unlike the pill wood-louse, cannot roll itself up into a ball, and is more or less helpless if it falls on its back as in the bottom centre photograph.

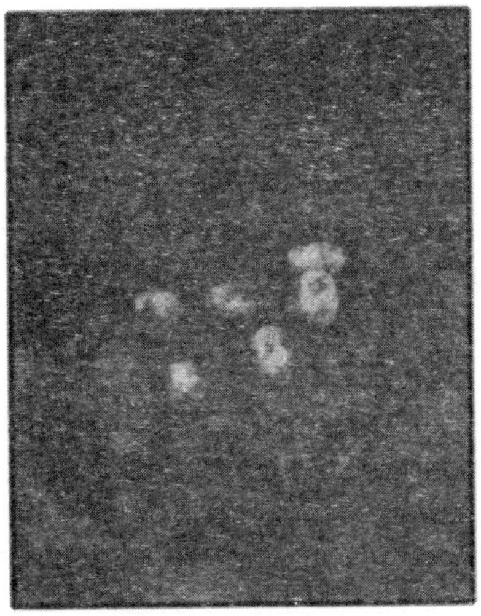

Eggs below ground

Larva resting in soil

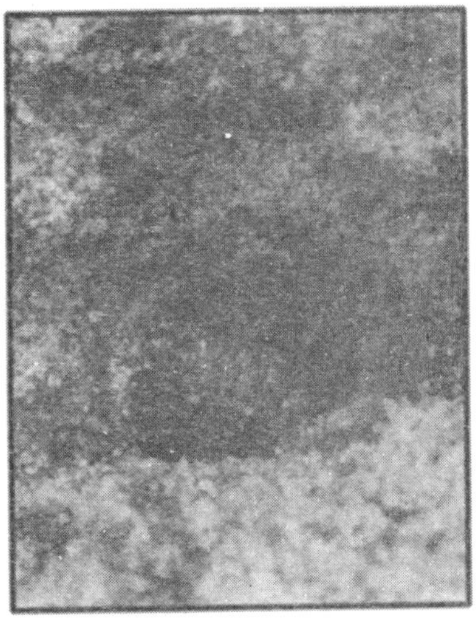

Excavating for pupating chamber

Pupa in soil

Beetle just emerged

Packing wings beneath elytra

Walking with head down

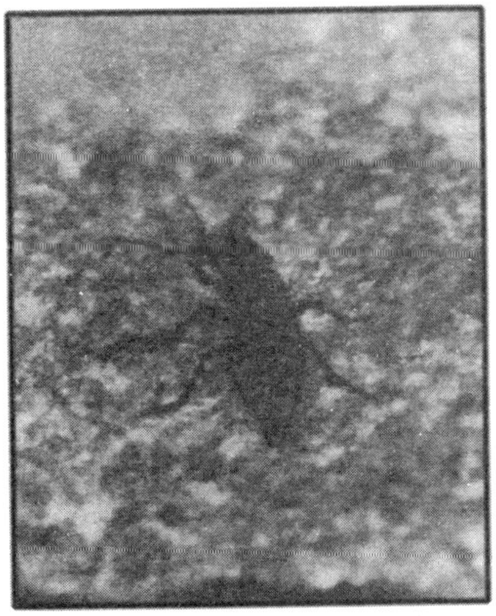

Back view of beetle

In fighting attitude

LIFE-STORY OF THE DEVIL'S COACH-HORSE BEETLE

Familiar in most parts of Britain is the beetle called the devil's-coach horse which has the habit, when disturbed, of bringing the top of the abdomen up in a curve over its head as though it were a scorpion. This is only " bluff," however. It is usually found in the soil or under stones, although it flies well and it belongs to the family of the rove beetles. It is a scavenger of both animal and vegetable matter. The eggs are laid in the ground, where also pupation takes place. The above photographs, by J. J. Ward, show the beetle's life-story.

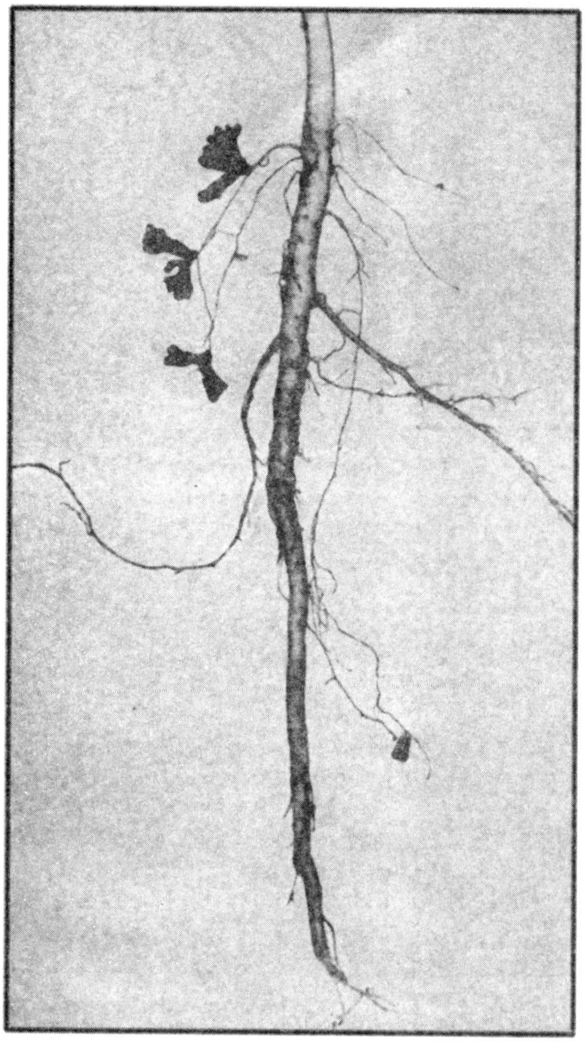

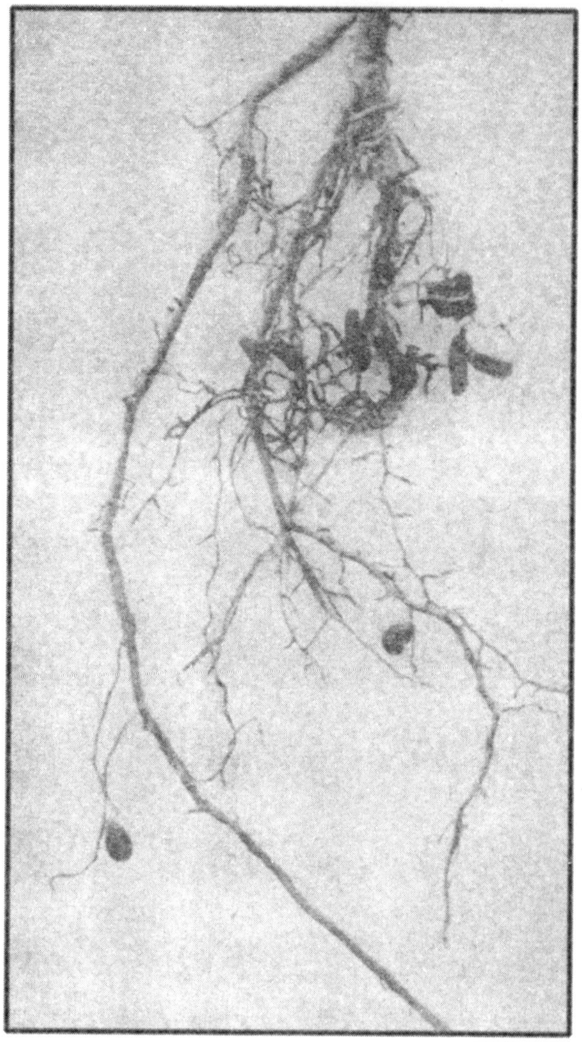

J. J. Ward

PLANT ROOTS WITH NODULES OCCUPIED BY BENEFICIAL BACTERIA

Enormous activity is for ever going on in the soil and the energy for this is derived, ultimately, from the sun. But animal life forms only a portion of the manifestations of this activity. Bacteria play the larger part. The bacteria belong to the plant family and consist of single cells which look like lumps of jelly Certain of them cause disease to animals, while others penetrate the fibrous roots of plants such as those seen above, and, taking up nitrogen from the atmosphere, pass on beneficial compounds to the plant and so provide food for animals.

the soil water. They multiply by division, one unit splitting into two or more. These grow and in turn divide again. The process of division may occur several times in 24 hours, so the increase in numbers may become very great in a short time, much greater than most people imagine.

There are two methods of counting soil bacteria. The older one has been in use for a long time, and most of our present knowledge of the different forms in the soil and their properties has been obtained by its aid. The principle of the method is to distribute evenly by thorough shaking, a definite weight of soil in a much larger volume of water, and then to add a small, measured quantity of this mixture to a gelatin (or other jelly) solution in which certain nutrient chemicals have been included. For convenience the gelatin is poured into flat glass dishes. A glass cover is placed over each dish and the dishes are put into a warm incubator. In a few days small spots appear on the gelatin. These are colonies of bacteria that have developed from the individual original ones in the liquid that was poured over the gelatin. The number of these colonies is counted, and, as we know what fraction of the original weight of soil is present in the gelatin, it is a matter of simple arithmetic to ascertain how many bacteria there were in the original soil. Many precautions have to be taken in this method. All vessels and dishes and materials have to be sterilised before use, to destroy any bacteria present, as otherwise these would develop into colonies on the gelatin and vitiate the results and the whole experiment.

THE method has many disadvantages. Unless the bacteria have all been separated from each other in shaking the soil with water, a cluster or colony,

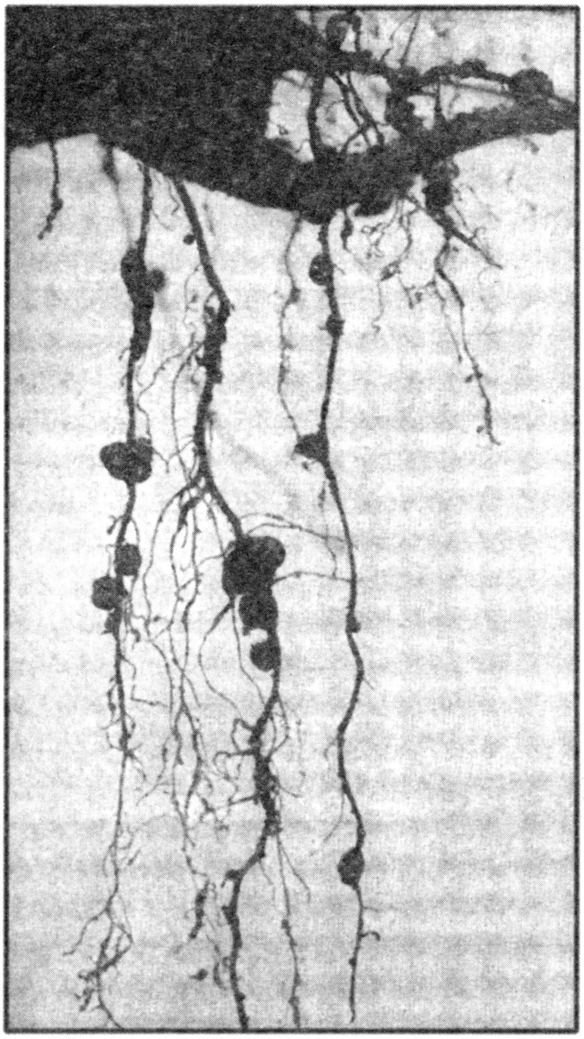

J. J. Ward

ACTION OF BENEFICIAL BACTERIA AND HARMFUL FUNGUS ON PLANTS USED AS ANIMAL FOODS

The left-hand photograph shows the root tubercle of a bean with nodules or knots formed by the action of bacteria. Incidentally the size of these minute organisms may be judged by the fact that if one of them were enlarged to the size of, say, a pin's head, an earth worm only two inches long would, if enlarged to the same scale, be about the size of a Tube train. The effect of the bacteria on the plant is explained in the opposite page. The right-hand photograph shows wart disease of potatoes due to a parasitic fungus.

containing very many individuals, will count only as one on the gelatin plate. Again, not all of the numerous different kinds of bacteria will grow in any one nutrient solution, so that many species, known from other experiments to exist in abundance in the soil, will be missed entirely in this method of counting. Indeed, the newer method of counting, that depends on staining the bacteria with certain dyes to render them easily visible under the microscope where they can be directly counted, shows that the older method accounts for only one-hundredth of the total number in the soil. But it has one great advantage ; the colonies are living and not killed, as in the staining method. A colony can therefore be picked out, transferred to another vessel and experimented on in the laboratory. Much of our knowledge of soil bacteria has been gained in this way.

In the soil the residues of plants and animals are broken down to simpler substances and eventually appear as a nitrate, which is taken up in water by growing plants and also by some organisms as a nutrient. If the soil is waterlogged so that air is deprived of entry certain species of bacteria attack any nitrate formed and reduce it to gaseous nitrogen that escapes into the air.

Other species of bacteria can use gaseous nitrogen directly and build it up into complex substances. This process was a puzzle for a long time, and the solution was quite striking. It was found that a bacillus living in the soil was able to penetrate the delicate root hairs of plants such as peas, beans, and lucerne, or alfalfa, which belong to the class of plants known as Leguminosae. It multiplied into a colony inside the root, obtaining its food supply from the material of the plant sap. At the same time it possessed the remarkable power of taking up gaseous nitrogen from the atmosphere and building complex nitrogen compounds from it These were passed

are broken down, and so on. The energy necessary to keep this process in continuous motion comes ultimately from the sun.

Besides taking a prominent part in this cycle of events, bacterial life itself follows in general a cyclic periodicity. If conditions are favourable they multiply, but not indefinitely. Eventually there comes a point when the immediate food supply is exhausted, or when the waste products of their own growth exert a poisonous effect. They may then pass into a spore, or resting stage, and remain dormant until more favourable conditions occur. Drying of the soil is a frequent cause of spore formation, and the dust raised by winds contains many such spores that can develop when they fall on to a suitable medium. As is well known, epidemics of infectious diseases are quite often caused in this manner.

Although for convenience in description the general action and life-history of bacteria has been described

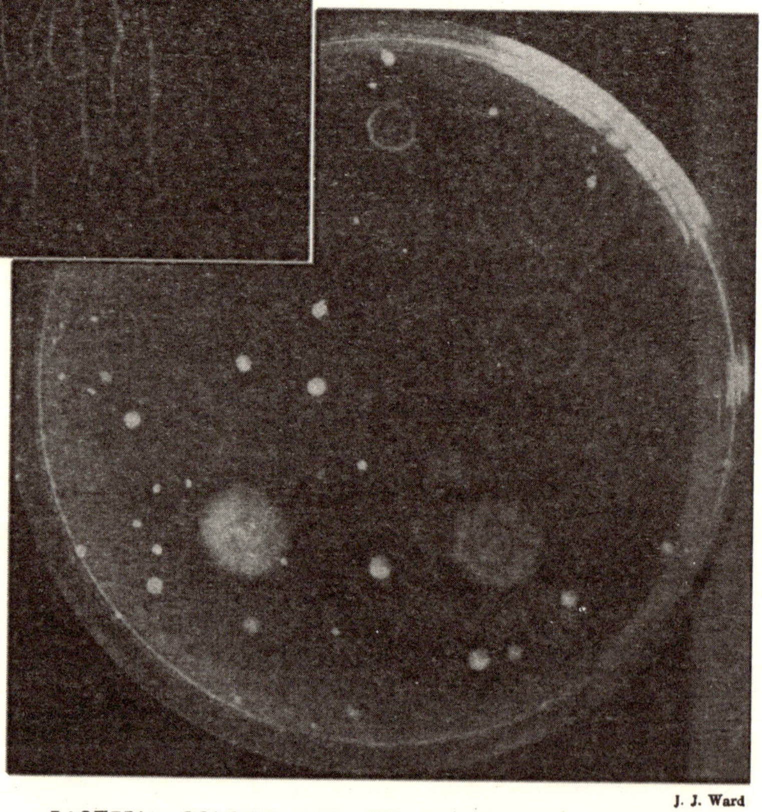

J. J. Ward

BACTERIA COLONY AND ITS BENEFIT TO PLANTS

Nowadays the scientist and the agriculturist go hand in hand to work. The lower illustration shows us a glass slide on which a colony of soil bacteria has grown, while above are two lucerne plants, the one on the left in its natural state and the other, much larger, after it has benefited by bacterial inoculation.

on into the plant. Thus both plant and bacteria supplied the wants of each other. The nodules seen on the roots of peas and beans mark the positions of these beneficial bacterial parasites.

Now that the life-history of this bacillus is properly understood it has been found possible by suitable methods of bacterial inoculation either of the seed or the soil itself, to get satisfactory growth of leguminous plants in regions where they were formerly grown either with difficulty or not at all.

THE action of bacteria in the soil may be summarised by saying that they take part in a great cyclic change in Nature. complex plant residues are broken down into simpler substances and these in turn are built up again by the next generation of plants and organisms into complex substances; these again in turn by itself, it does not constitute a distinct, walled-off part of the soil population. It is as closely dovetailed into, and affected by the other forms of soil life as is, for instance, the coal industry with the railways. It is especially bound up with the protozoa, for these organisms, although a very low form of life, are definitely to be classed as animals, and the majority feed on bacteria. The method of counting them is in its essentials similar to that used for bacteria, but it is possible to count separately those in the active living form and those that are dormant, or in the cystic stage. The numbers found are very much lower than is the case with bacteria, and they are, of course, much larger in size. The total number of all forms of protozoa in a saltspoon of soil rarely exceeds about a million.

A remarkable relationship has been discovered between some forms of protozoa and the bacterial members. When the latter decrease

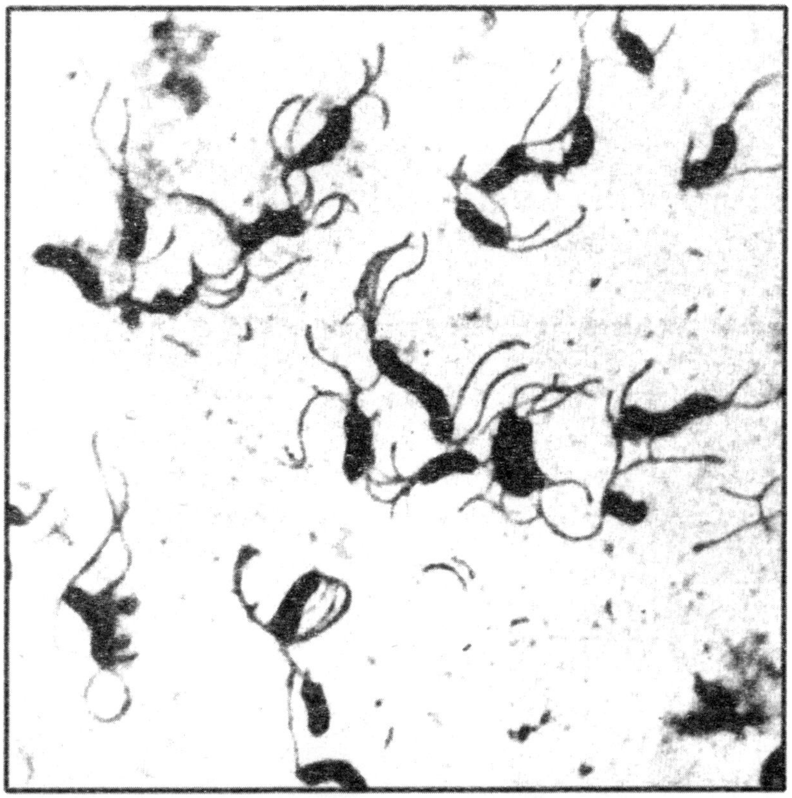

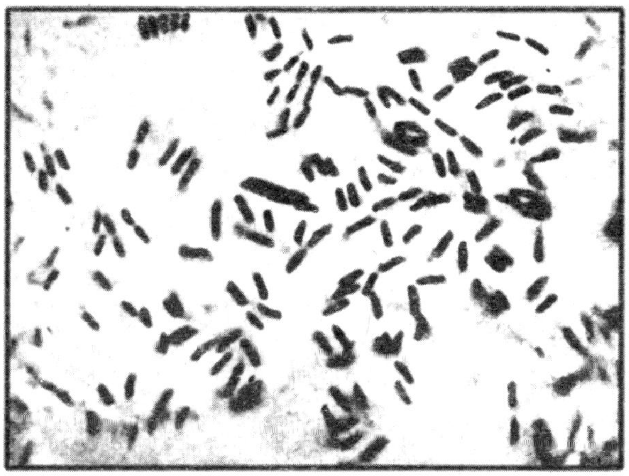

J. J. Ward

MICRO-ORGANISMS WHICH HELP THE PLANTS

Below are some bacteria which take up nitrogen from the air and supply the vital commodity to plants as explained in page 838. Above we see another photograph of micro-organisms of the soil. Both these subjects are greatly enlarged.

in numbers the former increase, and vice versa. The explanation of this alternating rhythm between these two groups of organisms is to be found in the fact that the protozoa feed on the soil bacteria.

Another important group of the soil population is the fungi. These are low forms of plants, and include a large number of different kinds. The larger forms, those that grow, for example, on decaying wood, and the mushrooms and toadstools, are familiar to everyone, but in addition there is a host of smaller kinds that can be found in the soil

Efforts to count these numbers by some reliable method have met with great difficulties, so at present no very definite idea can be given. The number seems, however, to be at least as great as that for protozoa.

Studies of their share in the activities of the soil population show that fungi are also concerned in the cyclic process of building up and degradation of the complex compounds, to which reference has already been made. Undoubtedly, the main interest and importance of fungi lies in the numerous plant diseases for which they are responsible, against which the farmer and gardener wage constant warfare. The great trouble is, to put it crudely, that the fungi are up to all kinds of tricks. A species that has lived a useful existence in the soil may, for some obscure reason, become parasitic on plants of economic importance, or again, one that has been the cause of a relatively minor plant disease in a small area may assume a virulent phase over a wide tract of country. Conversely, cases are known where serious plant diseases due to parasitic fungi have, for a time at all events, practically disappeared. This power of adaptability of fungi makes the task of the investigator very difficult indeed, and our present knowledge is not yet as extensive as in the case of bacteria or protozoa.

It is interesting to speculate how far the manifold activities of the whole soil population can be controlled by man. Is it possible to exercise a directive influence over their actions comparable to that already achieved both for yield and quality of crops

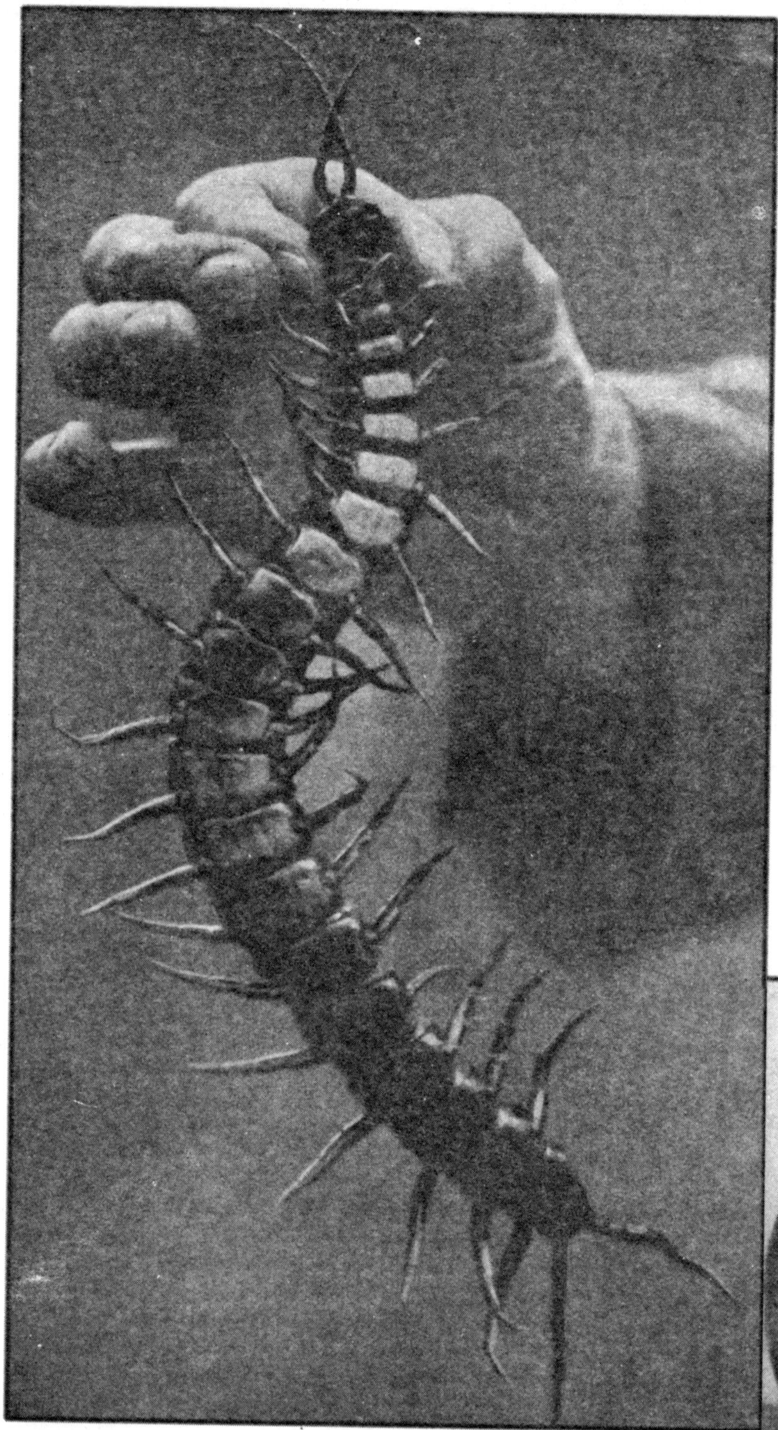

by using artificial fertilisers? Up to the present only the crudest measures of control have been achieved, and yet they have had immediate applications of practical importance. The inoculation of leguminous plants has already been mentioned, and this has been successfully used by farmers, particularly for lucerne, an important fodder crop.

Another discovery of great economic value was the method of curing "sickness" in the rich soils used for growing tomatoes and cucumbers in glasshouses, an industry of considerable importance near large towns. It was found that the extensive growth of plant pests of all kinds in the soil could be checked by a preliminary gentle sterilisation of the soil by heat or suitable chemicals. The pests were killed, but not all of the bacteria, and hence the latter were able to multiply unchecked and produce plant food in abundance from the rich stores of organic material in the soil.

A third application arose from studies of the organisms causing decomposition of straw and other litter used as the basis of farmyard and stable manure. When their action was understood it was found possible to rot down this vegetable matter directly. The process has found practical application in two directions. The waste vegetable litter, grass clippings, leaves, etc., in the garden can now be turned into an organic manure. The second direction is of especial interest to parts of the British Empire overseas, where there is much waste vegetable material, such as the straw of the wheat crop, or the crushed stalks of sugar-cane. This material can now be rotted down and returned to the soil as valuable organic manure.

MILLIPEDE AND CENTIPEDE

Both the names millipede and centipede, meaning a thousand and a hundred-footed, are of course only arbitrary. Below we see one of the former, rolled up into a protective ball, while above is a giant centipede. The tropical kinds are deadly poisonous.

Martin Duncan

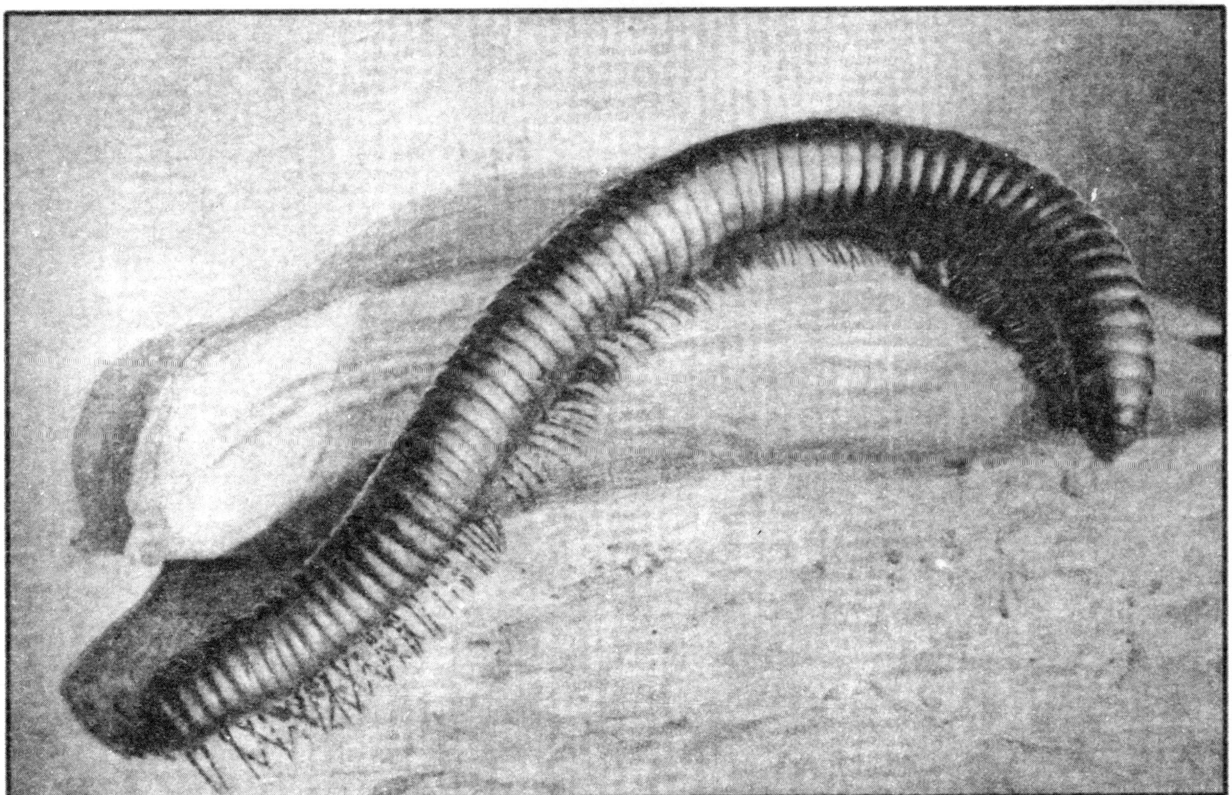

W. S. Berridge

GIANT MILLIPEDE AND A CENTIPEDE FROM TRINIDAD

The millipede is a sluggish creature moving about always on the ground and keeping out of bright light wherever possible. For defence millipedes are equipped with a number of so-called stink glands, which excrete a foul-smelling liquid to discourage attackers. The giant centipedes of the tropics, however (top), have the first pair of legs modified and extended forward to form poison glands. They are deadly creatures, these big centipedes, and deservedly dreaded in the hot countries where they live.

E.N.A.

BEAR LOOKS FOR A NEW POINT OF VIEW IN YELLOWSTONE PARK

Bears have a name for inquisitiveness as well in the wild as in captivity. This photograph shows one in the great Yellowstone National Park, U.S.A., which has climbed a tree to see if there is anything of interest up there and to view the landscape. The camera has caught him in rather a remarkable posture. Curiosity suggests intelligence and, indeed, the bear is one of the most clever of the larger animals. One reason for this, or perhaps a result of this, is the fact that it is an omnivorous feeder and always willing to try some new dish.

Chapter LXXVII

The Curiosity of Animals

By Frances Pitt

Author of "Wild Creatures of Garden and Hedgerow"

We are told that "curiosity killed the cat, but I doubt it, for it takes more than curiosity to kill a cat! Nevertheless it has killed many a bird and beast, from the inquisitive stoat, that must have a second look at anything strange, to the cunning inquiring jackdaw that must peer into every nook and corner, crack and crevice. Yes, curiosity is a great factor among animals, whether wild or domestic, and the more intelligent they are the more inquisitive they become about anything that is strange or unusual.

I was once watching bird life at the edge of a wood, a tit clan busy in the trees overhead tumbling like fairy acrobats at the extremities of the twigs, thrushes and blackbirds hopping to and fro, and a handsome jay preening its beautiful plumage against the dark foliage of a yew tree, when on the ground beneath the tree a slight movement caught my eye Something had stirred. I gazed at the spot and the "something" took shape and form, and became a stoat. It was a small female stoat, and she was staring in my direction, her little sharp face and pricked ears betokening keen interest in the strange object. I kept perfectly still, and she came into the open, leaving the hole which had given her shelter, so that I could note her slender elegant form and dainty cream chest. She sat up, in a begging attitude, her furry forepaws folded together, and gazed at me more earnestly than before.

F. R. P. Stringer

THE SOW WONDERS WHO

This remarkable and most entertaining photograph has caught a sow, kept in an outhouse, just as she leant over the half-door to see who it was that passed and if there was any prospect of food. The ears, the whole attitude in fact indicate a mild inquisitiveness rather than mere vulgar curiosity.

Did I make an inadvertent movement? I do not know. All I can say is that she suddenly bolted, turning like the flick of a whip lash, and vanished into her hole. Had she gone for good? Not a bit of it: within a moment or two she was peeping out again. The insatiable curiosity of the stoat tribe was compelling her to have another look. Alas! that curiosity is often fatal, for the watchful gamekeeper knows only too well that a stoat usually comes for a second look, so if he sees one, and it disappears, he has but to wait for it to reappear, when the little hunter of the woodlands will probably be quieted for ever.

The female stoat I was watching had a full allowance of inquisitiveness; out she peeped, dodged back, and then, when I made a little squealing noise, like a stricken rabbit, sprang forth into the open. She could not make me out, she was suspicious, and her suspicions deepened. She gave a hiss, flicked her dark-tipped tail, and vanished yet again, this time quite for good.

I have seen a weasel, it was on a wind-swept fjeld-side in Norway behave just the same. It had been hunting lemmings in and out around some boulders, but when it saw me it ceased its pursuit to have a better look at the intruder. Out it hopped, sat up, stared, dodged down a hole,

TIGER CAUTIOUSLY CURIOUS

In this case the weather was very cold and the tiger, an animal coming as it does from a hot climate was wondering whether the temperature was bearable outside or not. The photograph was taken at the London Zoological Society's Gardens and gives us a splendid and typical tiger attitude.

and came out again, quite half a dozen times before it went on with its business.

Now the beautiful pine marten, a relative of the weasels, which in appearance combines their lithe grace with the attractiveness of a cat, is also desperately inquisitive. I had a tame one once, known as " The Mart," the most lovely and confiding of pets, and any thing strange had to be investigated at once. She was not happy until she knew all about it, whether it was a new ball to play with, or merely that I had got a new coat. The latter must be examined, especially its pockets, but she was not, after all, quite so inquiring as my two otters.

The otter also belongs to the weasel tribe, and in it the inquisitiveness of the family seems to reach its climax, but this curiosity is tempered with a wise discretion. My two otters, Thomas Romeo and Madame Moses, are of the most inquiring turn of mind, yet deeply suspicious of anything strange. If the new thing is quite outside their experience they give it a wide berth until satisfied it is harmless, after which they paw it and nose it all over. This at times is embarrassing, especially when one is trying to take photographs of them, as both otters know all about cameras, yet are never tired of examining them afresh. Hence the difficulty, for one cannot get a good photograph of an animal that is poking a wet and muddy nose against the lens.

One of the most striking instances of this inquisitiveness was when the two otters were paying a visit to the house, and someone opened a door which was seldom used. They immediately ran to it, stood up on their hind legs and examined the hinge as high as they could reach, just as if they were trying to discover how, where and why, it had opened.

Now, curious as a cat undoubtedly is concerning anything unusual or out of the common, I have never seen one show such inquisitiveness as Tom and Madame Moses. Still, I do not wish to decry the curiosity of puss! I have just seen a workman prop a ladder against the stable wall, and he had hardly left it before a cat came by, paused, looked at the ladder, and turned aside for purpose of a closer inspection. Her inspection included climbing the ladder to its topmost rung, which was just like a cat! We could multiply instances of cat curiosity by the hundred if not the thousand, for, after all, puss is liable to this little failing—if failing it be.

I consider inquisitiveness a sign of an active, inquiring mind. Without intelligence you cannot have curiosity. Yet the average dog, whose wits are certainly keen enough, does not betray anything approaching the curiosity of the cat, let alone an otter, but I think the reason is he is too preoccupied with his own affairs. The cat is, so to speak, a lady of leisure, with time on her hands, and able at any instant to turn aside for an interesting investigation. The dog's attitude to the world and life in general is rather that of the business man off to the city; he must get on with his job, even if it is only that of following his master, and is not lightly distracted or turned aside. Pure curiosity he leaves to the cat.

Birds have their share of it, too. One of the prettiest cases of inquisitiveness I have seen was when a party of blue tits investigated a waiting motor. The car had been left on the drive in front of my home, and the tits were in the habit of coming to feed upon a coconut hung by one of the windows. Looking out of the window I was amused to see some half dozen blue tits about the car. One little bird was sitting on the polished metal top of the radiator, pecking at it, as if aiming blows at its own reflection in the plate—was it really having a fight with the tit that looked out of the silvery surface? At any rate it appeared like it. But the rest of the party were simply investigating; two hopped to and fro across the top of the hood, peeping and peering into every fold of the material, another explored the running board, and a fourth peeped in at the window. More inquisitive little birds I have never seen.

The domestic fowl has its share of curiosity, as was well shown one morning when I took a tame fox through the farmyard. The numerous cocks and hens caught sight of the vixen at once. They began to chuckle and cackle in great excitement, but instead of running away they ran after her. The barn-yard ducks joined the throng to quack loudly at the sight of the strange little red dog, and follow the fox and me as I led her homeward. Far from showing any instinctive fear of their

INQUISITIVE BUT NOT OVER-BOLD

Here is a lioness of the London Zoo investigating conditions from the refuge of her indoor cage. The ability to know without being known is of vital importance to those who get their living in the wild from the hunting of wild things, and caution must mingle with curiosity as a condition of existence.

F. Pitt

OTTER INVESTIGATES A CAMERA AND AN ANTELOPE A TIN OF FOOD

The lower photographs show a quaint instance of animal curiosity. Two otters, pets of the writer of this chapter, found this strange implement, a camera, one day. It was to be used for photographing one of them, but the otter was so interested in the tripod, the lens, and in fact, the whole instrument, that another camera was fetched and these happy "snapshots" obtained. Above are antelopes from a herd preserved in Wichita National Forest, Oklahoma, investigating food supplies.

RACOON, PROVIDER OF A POPULAR KIND OF FUR, ON THE LOOK OUT

Tree tops are the haunt of the racoon, which usually lives in a hole in the trunk high up from the ground and thus has home and look-out tower in one. Curiosity is, then, a special feature of the racoon's existence, for the animal is constantly watching from some lofty point of vantage for "something to turn up" whether it be danger or the chance of a meal. The racoon is a small relation of the bear, its body usually measuring about two feet in length and, by day, it usually keeps to its tree, descending at night to feed.

SEALYHAMS AND ALSATIANS INTERESTED IN PHOTOGRAPHY

Dogs are not, as a rule, so pervaded by the sense of curiosity as their chief companions in domesticity—cats. The dog is usually concerned with its master and the activities connected with him, and what is called a well-trained dog will subordinate the satisfying of its curiosity to the master's interest at the word of command. The cat, on the other hand is at its friendliest a guest, not a servant. Below are some Sealyhams and above an Alsatian and her pups " registering " (as they say in the cinema world) curiosity.

F. W. Bond

Mondiale

BEARS REACTING TO THE UNUSUAL IN TYPICAL BEAR FASHION

In that wonderful sanctuary for wild animal life, the great Yellowstone National Park, occupying some 3,350 square miles of territory in the states of Wyoming, Montana and Idaho, U.S.A., animal and man have begun to understand each other a little better and, consequently, some strange sights are sometimes witnessed there. In the lower photograph, for instance, we have a bear examining a, to him, strange new creature, the motor car. Above is a polar bear, which has just arrived at a Zoo, taking a long, wondering look at its new home

WILD LIFE IN THE TRANSVAAL ENCOUNTERS THE CAMERA

A camera, especially if it be standing upon a tripod, is such a curiosity-compelling object that few wild animals can resist it. Fear of the unknown and the itch to know something new—for there is, perhaps, a certain sameness even in the fulfilling of the law of "eat or be eaten"—struggle, each emotion with the other, until curiosity finally wins. Below is a giraffe daring to satisfy its inquisitiveness, and above, kudu at a waterhole where a photographer was lying in wait. Waterholes are favourite "pitches" for the wild animal photographer.

E.N.A

Neville Kingston

OTTER AND THE FIERCE TIGRELLO, OR MEXICAN WILD CAT

Pausing as it swims, the otter (bottom) distracted by some sound or sight on the bank tilts up its head to observe. At all times the otter is, like the famous elephant's child in the "Just So Stories," filled with insatiable curiosity and when the animal is kept as a pet, this quality, which is developed a hundredfold by a domestic environment, affords endless diversion. The upper illustration gives us a glimpse of a tigrello, or Mexican wild cat, stationed in the fork of a tree and looking for the cause of the disturbance it has noticed.

MASS CURIOSITY AMONG THE MONKEYS
This photograph was taken at the Washington Park Zoo, and shows a great number of monkeys, the most inquisitive of all animals, climbing a tree after their leader, who suddenly conceived a desire to examine the prospect from the top.

ancient enemy the hens seemed filled with curiosity. The country folk have a legend to the effect that when a fox finds fowls roosting up in a tree, instead of being safely shut up in their fowl-pen, he gets them to come down and come within his reach by the following device. They say that he takes his brush in his mouth and spins round and round in a crazy dance, which so mesmerises the hens that they fly down into his jaws. This is, of course, simply "an old woman's tale," but it has a slight foundation of fact, namely, that fowls are so inquisitive that they are quite likely to fly down from their roosting-place to get a better look at a strange animal.

In the "days of old" the curiosity of the mallard was turned to good—or should one say evil ?—account. Certainly the result was bad from the ducks' point of view, however excellent from the standpoint of those who ate the mallard! I am referring to the decoys which were then in common use in many parts of Europe, and of which one or two still survive in working order in Great Britain. The success of a decoy depends entirely on the inquisitiveness of the ducks that are to be decoyed.

Apropos of the fascination my pet fox had for the fowls, she was equally attractive to the tame mallards. They would come ashore from the pool when she was at play in the garden, and follow her about to watch her antics. Sometimes the vixen made rushes at them, and I feared would catch one, but they did not seem to have any apprehension of danger, and, after scattering before her rush, soon waddled back to have another look.

However, it is not only mallard, but many other kinds of duck that are troubled by an inordinate curiosity. I have seen teal, widgeon and tufted ducks much excited at the sudden appearance of a cat, following it as it ran along the waterside as far as they could swim after it.

In the case of many wild animals, curiosity is neutralised by fear and caution. Though the strange thing is interesting it may be dangerous, and the impulse to have a look at it is overcome, or partially overcome, by the equally strong impulse to give it a suitably wide berth.

I have often noted this when trailing wild creatures in the snow, their tracks betraying how they have cautiously circled around anything new and strange, before venturing near enough to investigate it.

It is the same with domestic animals, such as cattle and horses. You will see in them the same conflict of emotions, though in their case curiosity nearly always wins, for they have not the same deep fund of timidity which makes the wild animal so nervous.

Once a bird or beast knows all about a thing, it loses interest in it. What is well known is dull and not worth troubling about, so the creature goes off to seek something new. Just as inquiry is the salt of existence for us, so it is for the lower animals. Curiosity is nearly as great a force among them as it is with human beings, and that is saying a great deal. Curiosity shows us mind at work, hence curiosity in the animal world means intelligence to the fore.

Chapter LXXVIII

The Marvel of the Paper-making Wasps

By John J. Ward

Author of "Some Nature Biographies"

It was a day in late autumn, and a lady called me into her garden to see a strange thing that was happening. A wasp had arranged a row of fat, white grubs on the gravel path, and at almost regular intervals was adding another to the line—an amazing incident, and one around which there hung a wonderful story.

Those white grubs in the natural order of things were destined to become wasps, and the worker wasp we now saw removing its charges from the nest, with its inborn instinct of straight rows of cells, was arranging these half-developed wasp grubs in proper cell order, side by side. Until that very hour that worker wasp had tenderly fed and nursed those same grubs, but now things had changed; it was ruthlessly dragging them from their cells and conveying them into the open to perish with cold, or to be devoured by an insectivorous bird.

Disaster had suddenly fallen upon a great wasp community of some sixty thousand inhabitants; order had given way to confusion, and only a few of the newly-emerged worker wasps imbued with the keen instinct of work, like our example on the gravel path, were keeping a cool head and trying to make the best of a dire calamity. The older workers had lost heart and rested dazed and confused amongst the cells, or around the outside of the nest.

The trouble began the night before, when the temperature suddenly lowered until by early morning it had reached a stage of frost—the death knell of the wasps. Notwithstanding that the wasps were all safely in the nest below ground, so sensitive were they to what had happened that all hopes for the future had vanished completely. Unlike the honey bees, they store no food in their combs, and there is no shelter for the winter in their frail home.

Later in the morning when the sun had broken through, the drowsy and half-paralysed wasps developed some amount of activity, but since there was now no hope of rearing the thousands of grubs in the cells, the wasp on the gravel path, and a few of her sisters proceeded to remove their charges from their cells to perish quickly in the open. "How merciful, rather than to let them die slowly," says the moralist, but there is no mercy involved in this action of the worker wasps. It is a last instinctive act for the welfare of the race; the individual may be sacrificed, but the race must be maintained. This last act of removal of the grubs is purely one of leaving the nest in a good sanitary condition in readiness for the emergence of the males and young queens on which depends the destiny of the next generation. Males and young queens will continue to emerge from the large closed cells in the combs after the workers that built those cells have perished.

The male wasps do not appear until late summer, and they await the advent of the young queens. They do not play any part in the building of the nest or rearing of the wasp grubs, and during the day leave the nest and sun themselves on the leaves of trees such as those of the sycamore, where they are fond of licking off the sugary substance known as "honey-dew," an exudation of the aphides. They are not provided with stings, and they can easily be caught by the hand if you are quite sure of your identification, their characteristic features being a long, slim body and longer feelers than those of the worker wasps.

When the young queens begin to emerge the waiting males immediately find their mates, for, unlike the honey bees, the mating not infrequently occurs just outside, or even on, the nest, and with individuals of their own great family. It is somewhat astonishing, too, how each of the couples in re-crossing each other's tracks during their courtship seem to concentrate entirely on their selected partners.

After mating the male wasps die off, and each young queen at once seeks some warm quarter in a sheltered spot in the hollow of an old tree, or beneath a thatched roof or some similar situation, to hibernate for the winter. When, in late October, she has found her winter residence, she settles comfortably down, first gripping firmly with her mandibles some projecting particle of bark, or, in a roof, perhaps a protruding straw, and then folds her wings and tucks her legs close to her body. Suspended by her strong mandibles she sleeps away the winter months, awakening some day in early May. From this it follows that any wasp we may meet with between November and early May will be a queen wasp, and the potential mother of a new wasp community; hence the killing of that wasp means one wasps' nest less for the following summer. As that nest may contain from twenty to sixty thousand individuals it is obvious that the proper time to kill wasps is early in the year.

While the structure of the forsaken nest holds together it becomes the prey of slugs, snails, wood-lice, spiders, earwigs, and other creatures, and that wonderful fabric so marvellously built by the many thousands of workers soon falls to ruin. Here, again, the wasp differs from the honey bee whose hive shelters the colony for the winter period; and although our sympathy may lean towards the wasps whose labours, at the very zenith of their success, were so suddenly doomed, yet we have but to imagine the consequences if it were not so. If the queen alone can become the mother of sixty thousand offspring from early May until she dies of old age in October, we can only wonder what would happen

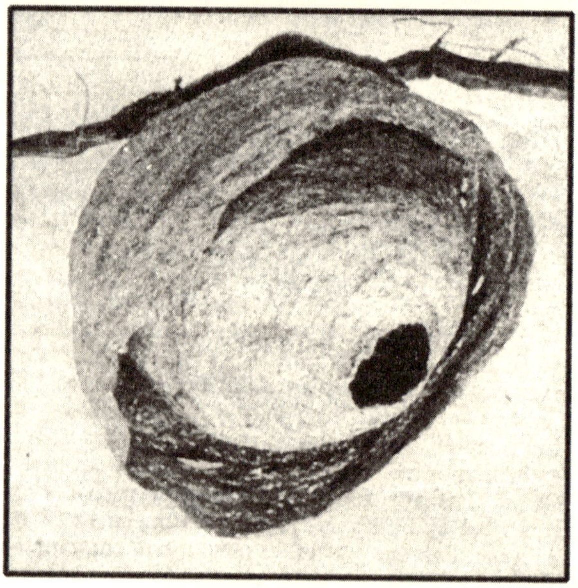

TREE-WASP'S NESTS H. Bastin

As opposed to the ground-wasps, which make their nests underground, the tree wasps construct theirs in the open air. Below is a nest in the earlier stages of its building, and above one found in a packing case which provided a handy source of "wasp paper."

if the community recommenced in full strength in early spring; there would, indeed, be a plague upon the land, and wasps would reign supreme.

There are two kinds of social wasps, tree-wasps and ground-wasps. Both kinds construct their nests by much the same methods, but the former build their nests in the open air usually amongst the branches of some prickly shrub, such as a holly bush, or a hawthorn hedge, although one species shows a partiality for the branches of a fir tree; while the ground-wasps excavate the soil and build their nests in darkness underground, and occasionally under a roof, or even in a hollow tree Hornets, which are merely a large species of wasp, usually prefer the two last-mentioned sites. Whether it be tree-wasp, ground-wasp, or hornet, with each kind the building is commenced in the spring by the queen after her hibernation. There are three British species of tree-wasps, and, including the hornet, five ground-building species, although one of these, the Austrian wasp (*Vespa austriaca*), is apparently a sort of "cuckoo," or interloper, in the nest of the red wasp (*V. rufa*), and appears to be represented only by males and fully-developed females, the workers of the host wasp rearing its family.

The queens of the two common ground-wasps (*V. vulgaris* and *V. germanica*) commence to build at the beginning of May. After awakening from her winter sleep each queen pays considerable attention to her toilet (for she and all her kind are models of cleanliness), after which she tours the neighbourhood in search of a nesting site. Buzzing gaily, and full of the joy of life, her attention is largely given to the bottoms of hedgerows, projecting tree roots appearing particularly attractive. After many unsuccessful examinations she at last becomes interested in a disused mouse run, in and out of which she continually travels with obvious excitement. It is certain, too, that she has decided on that spot, for she is carrying out in her strong mandibles bits of nut shells, tiny twigs, little stones, and other odds and ends. She is tidying up the place to adapt it for her home.

After an hour or more of this occupation she comes outside and, alighting on the bank, she devotes several minutes to brushing her face, head, and body briskly by means of her bristly legs, finally drawing each of her feelers from base to tip through her folded forelegs. Then all is ready, and, taking to her wings, she circles several times around the entrance to the burrow, each circle getting wider until she is well up in the air, when she puts on speed and makes a definite course. During that circular tour round the site of her nesting quarters her mental faculties had made a comprehensive record of details that, on her return, will guide her again to the exact spot.

If we could follow her in her flight we should discover her alighting at some distant spot to investigate some wooden palings, or an old tree, and the chances are that the wooden fence would prove most attractive to her. Alighting upon it, she would proceed to scrape it with her strong mandibles. Splinter after splinter she would tear off, munching up each one, and mixing with it saliva secreted from special glands with which her tongue is provided, until her mouth was full of paper pulp. Then she would fly rapidly back to her building site, encircling its entrance with diminishing circles as she approached and then rush in with obvious delight. Perhaps at the end of the run there would be a space about the size of a hen's egg, and in that space she would commence to build a nest that by autumn would probably be as large as a four-pound loaf.

On a bit of exposed root in the roof of the chamber, which before leaving she had nicely cut and cleaned

with her mandibles, she now by means of those same useful tools spreads her mouthful of wood pulp, forming it in the shape of a flat disk—the foundation stone of the new wasp building. Further visits to that same fence are then made by this industrious little creature. Each time she scrapes away as much wood as she can conveniently manage, and soon she has formed four open paper cells suspended mouth downwards from the disk, in each of which she then deposits an egg. More paper pulp is then brought, and presently the cells are covered with a kind of umbrella. Each time as she leaves the burrow she drops outside a piece of soil, for she is excavating as well as building. Other cells are then added around the first four, the base being extended to receive them. Such is the beginning of the first comb of cells, which, unlike the perpendicular combs of the honey bee, are arranged horizontally, while the horizontal

H. Bastin

INTERIOR AND EXTERIOR VIEWS OF A FINE NEST OF GROUND-WASPS

The lower photograph shows a nest with the outside layer of "paper" removed to show the plan of construction. The combs are built horizontally, not vertically, as is the case with bees, and further, no food is stored in their cells. Above is a photograph obtained by scraping away some of the earth from a bank in which some ground-wasps had hollowed out a cavity.

H Bastin

QUEEN WASP AT WORK IN HER WONDERFUL REALM OF PAPER

Queen wasps not only rule but found their city states. Each queen wanders away in search of a suitable nesting site and, having selected one, begins to manufacture " wasp paper " out of wood pulp and to construct the nucleus of the nest. She lays eggs in the first few cells, and as she goes on with her work these eventually hatch into workers, who assist her at the ever-increasing task of feeding the helpless grubs. Eventually the queen is set free from every duty but egg-laying. Below we see her attended by workers and above, laying eggs.

cells of the honey bee also differ from those of the wasp, the cells of the latter being set in the perpendicular position mouth downwards.

The queen wasp has not got very far with her work when the eggs laid in the first four cells hatch into wasp grubs. Then this so-called "queen," who is really a maid of all work, has to feed her offspring. She goes out to seek for aphides, tiny caterpillars, and similar tender meat for her hungry babies. With a mouthful of such food nicely minced she returns and feeds them from beneath as they hang suspended head downwards. They cannot fall from their cells, for they are firmly fastened to the base of their egg-shell. The grubs grow rapidly, and their busy mother in her spare moments, when not feeding them, continues to make more new cells, putting an egg in each one as it is completed to her satisfaction.

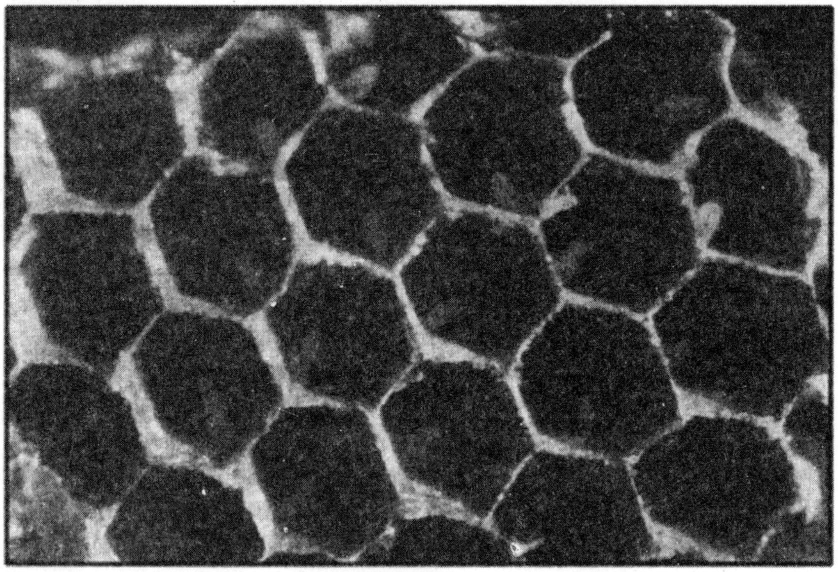

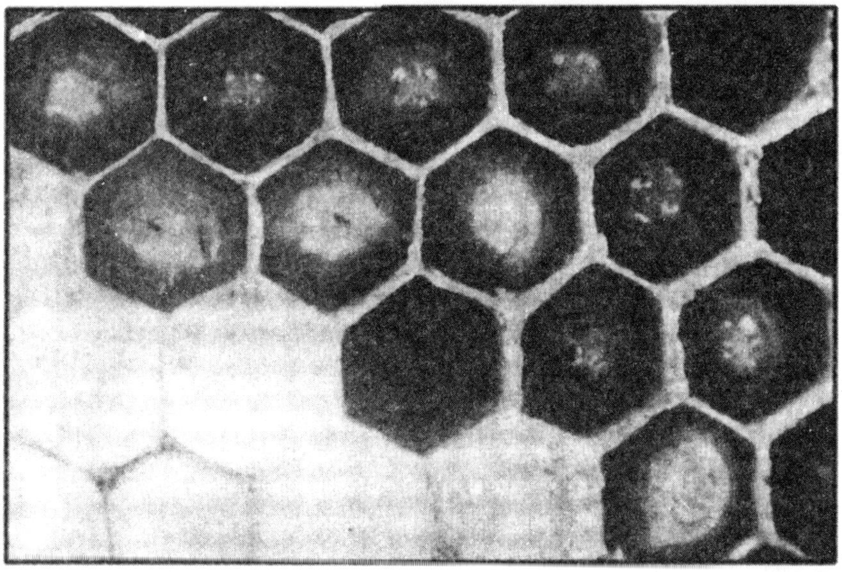

H. Bastin

GRUBS AND EGGS IN THE CELLS WHERE THEY GROW

Wasp eggs are, as they are laid, stuck to the side of the cells, one to each cell, at right angles to the cell wall. When the grubs have hatched and grown they, in time, seal up the mouth of their cells with a white, domed cap of silk and become pupae. Below are some cells, those to the left being capped, and the upper photograph shows some eggs as they are stuck to the cells. Both photographs are magnified.

Meanwhile the ravenous wasp grubs call for more and more food, while new baby grubs are continually hatching out from eggs in the later made cells. Nothing daunted, the mother wasp works harder and harder, and, in her quest for food, she boldly stalks unsuspecting flies, pouncing on to them while they are resting and rolling them over so that they suddenly find themselves a wing short, cut off by her sharp mandibles While the fly is discovering its injury its other wing is removed, and maybe its head; its body then being converted into minced meat for the wasp's hungry family at home. At this season sweet food is not needed: her children are carnivorous and they must have meat provender. A squashed caterpillar on the garden path will serve her equally well; she will remove it piece by piece; that, indeed, is the function of the wasp; she is first and foremost a very hungry scavenger of all forms of flesh food.

If by any chance she, or her workers that come later, should in their area of work enclose a butcher's shop, that is a glad event, for there flies can always be found without loss of time; and the butcher's board provides a splendid stalking ground. Sometimes in stalking a fly that happens to alight upon the meat the wasp's mandibles get a grip of the beef, and then the wasp makes a discovery—it finds that beef is better than flies. After that happening it returns many times, but for bits of beef and not for flies. That is a sad story, for the wasp's morals were led astray; that was not its natural function. Its intentions were of the best, but temptation was put in its way. It was still continuing its scavengering work in removing dead flesh food, although it is difficult to get the butcher to appreciate that point. We should remember, too, that by far the greater number of wasps are never fortunate enough to find a butcher's shop, or even a village, yet they successfully rear their families in the wild.

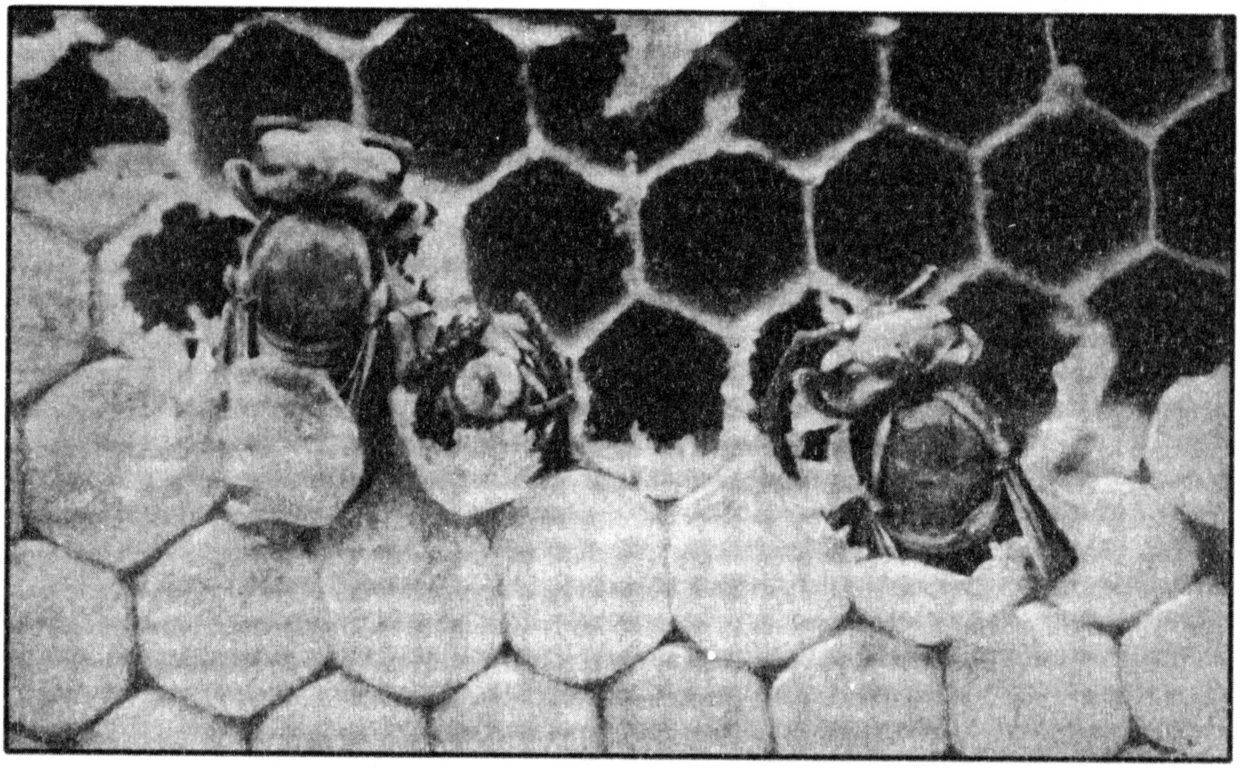

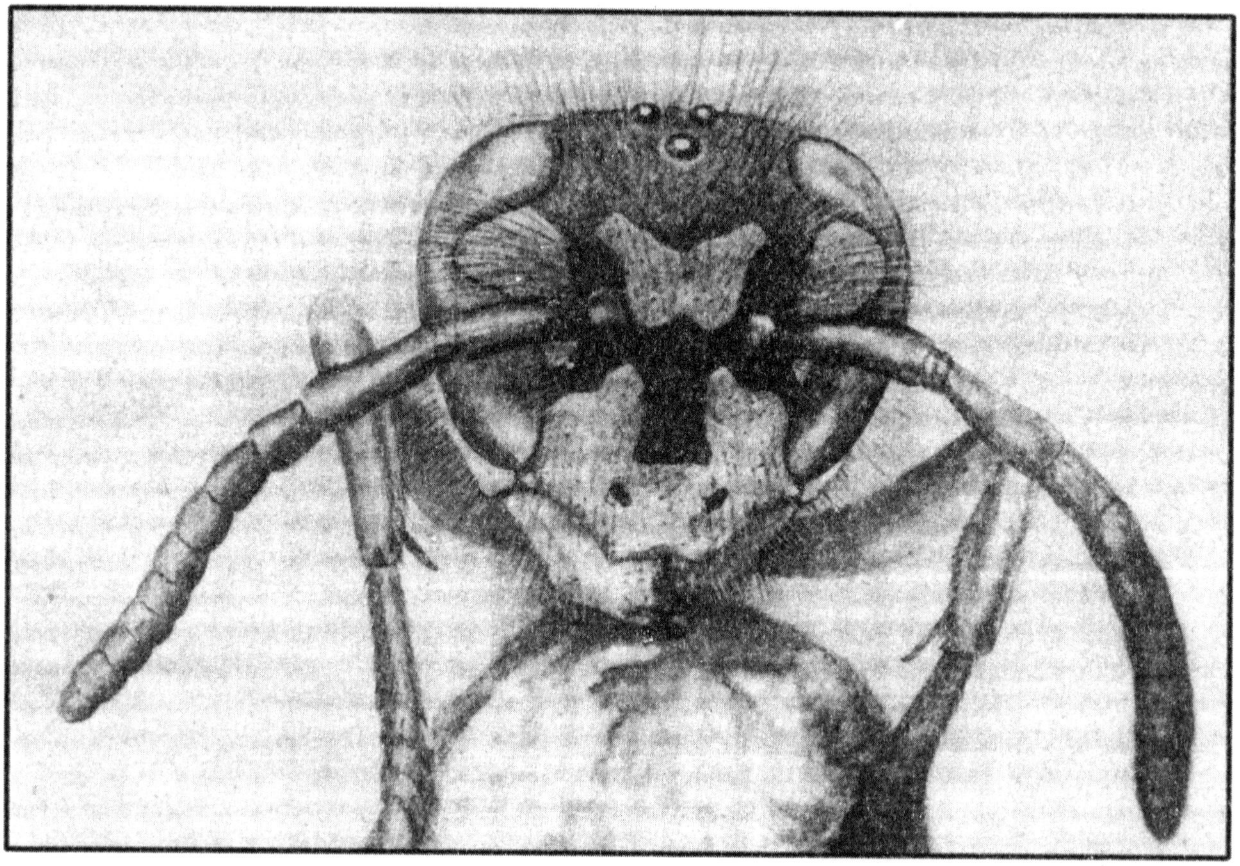

H. Bastin

WORKER WASP FACE TO FACE AND WORKERS EMERGING FROM THEIR CELLS

Face to face, a worker wasp looks like something from a story by Edgar Allan Poe when enlarged to the size we see it here. The insect was photographed as it sat on a twig, and the camera lens was placed so that the wasp looked right into it. The upper photograph shows part of a comb in a wasp's nest enlarged with some young workers just breaking their way out of the cells they have sealed for themselves. As can be seen by the empty cells above them, some of their fellow workers have already got out into the world.

H. Bastin

HEADS OF WASP AND HORNET, AND A QUEEN WASP HIBERNATING

When the queen wasp hibernates she seeks out some sheltered place, takes a firm grip with her jaws and folds her wings beneath her abdomen. The lower left hand photograph shows the underside of her head, with two pairs of legs and the antennae, while the upper photograph shows the whole insect clinging to a piece of rotten wood. She generally hangs tail downward. The lower right-hand illustration is of a hornet taken looking straight into the lens of the camera. These photographs are all greatly magnified.

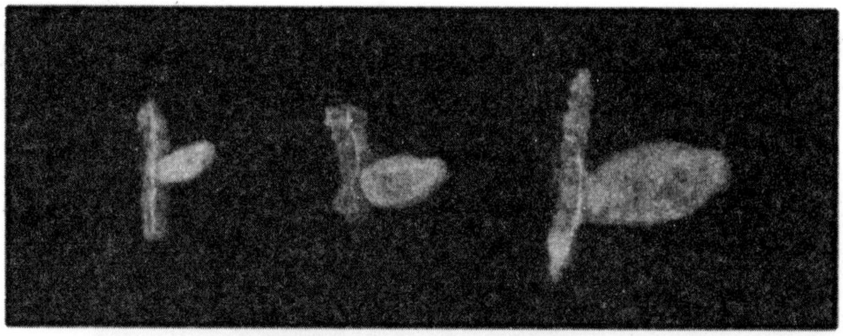

Three stages of the wasp's egg

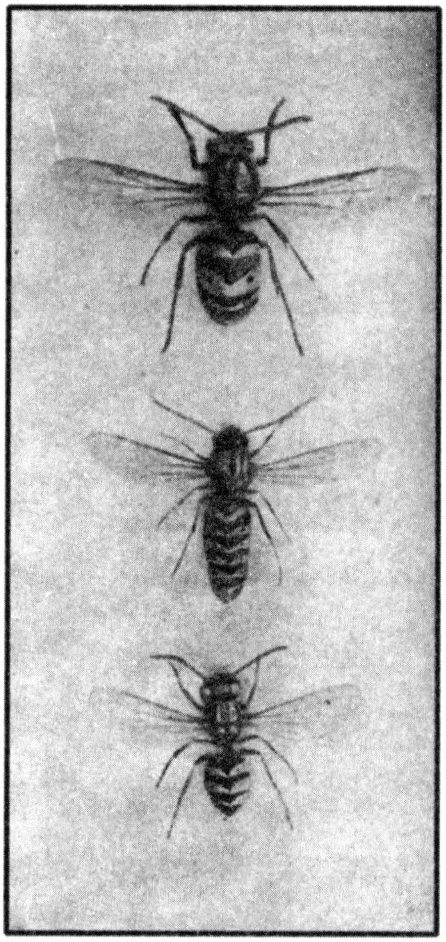

Queen (top), drone (middle), worker (bottom)

Wasp grubs at various stages

Full grown grub (left) and two wasp pupae

Point of wasp's sting and needle-point (right)

WASP'S LIFE FROM EGG TO IMAGO, AND THE PAINFUL STING

The three eggs seen in the top left photograph were left attached to the fragments of cell so that they should not be damaged. These illustrations, all magnified, give us the story of a worker wasp's life. The larvae (centre left), which cling to their cell by means of a "foot" at their hind end, transfer themselves a few days after hatching to the end of the cell, where they hang head downward. The imagos, or perfect insects, are also shown here, and a sting compared with a fine sewing needle. The photographs are by H. Bastin.

When some two dozen or so cells are formed the wasp grubs that first hatched out are full fed, and they themselves seal in the tops of their cells, secreting from their mouths a kind of liquid silk, which they spread on the interior of their cells and finally form a closing cap with it. The cell then becomes a cocoon for the development of the pupa stage. The mother wasp has no interest in the closed cells, her attention being fully occupied in feeding her increasing family. She exhibits a motherly instinct quite unknown to the queen honey bee.

Some ten or twelve days later the inmates of the first four closed cells have matured, when they bite their way out. That is, indeed, a glad day for the mother queen, for she then has four (sexually imperfect) daughters, all as able as herself to go out and capture food for her growing family. The advent of these first workers is but the beginning of continually increasing assistance, for they have scarcely got working before other workers emerge from their cells. Presently the mother wasp is entirely relieved of feeding her young, or at least from having to go out to seek for food. She is content to superintend generally in the nest, and to deposit eggs as fast as her daughters can build cells to receive them.

By the time that the first workers have appeared the queen has herself constructed a complete little nest with one comb of about two dozen cells, but with the advent of the increasing force of workers the cover of this nest is soon cut away for enlargement, and the comb area is likewise extended. Then a short strong pillar is lowered from the centre of the first comb, on the end of which three or four cells are built, other cells then being added outside these with further pillars to support them. In that manner a second floor, or comb, is constructed. From seven to ten combs are eventually formed, the largest at the centre containing from three to four thousand cells.

THE cells in the larger combs are used at least three times, for, as the worker wasps emerge, other workers tear off the remnants of the covering caps of the cells and clean them out in readiness for the queen to lay in them other eggs. So the tide of life spreads from the centre of the comb to the outer margin and then recommences again at the centre. Between the combs is just sufficient room for the workers to move about while tending the suspended wasp grubs.

As the nest grows a small army of the older workers have to become miners and excavate around the nest, carrying mouthful after mouthful of soil to the outside of the burrow.

When autumn approaches large cells are formed on the two lower combs, and these cells when closed have rounded instead of flat tops. They contain grubs which have had special feeding to produce queens and males—the parents of the colony for the following year. By this time the outer envelope of the nest may consist of eight or more sheets of wasp paper, every particle of which has been formed by the many thousands of workers, each one bringing in its mouthful of wood pulp.

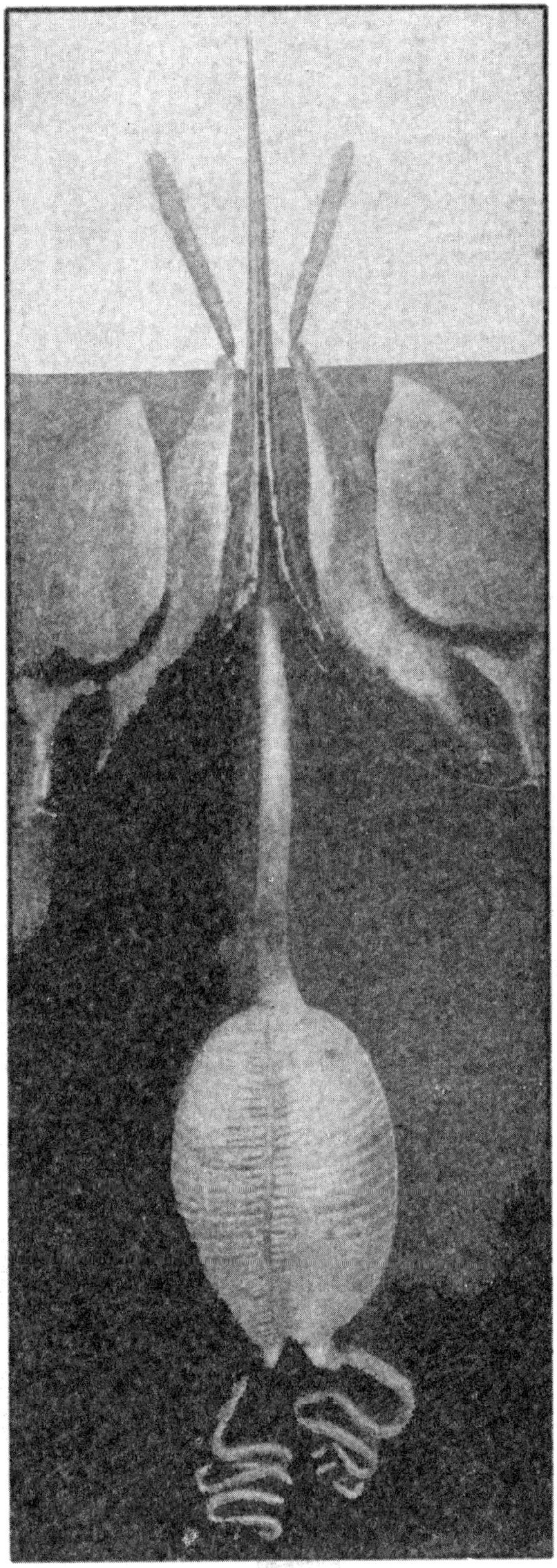

H. Bastin

MODEL OF WASP'S POISON APPARATUS

At the bottom are the poison glands communicating with the poison bag. This last is in turn joined by a tube to the hollow "director" the sting itself. Only the top of the sting and two of its special feelers are outside the wasp's body.

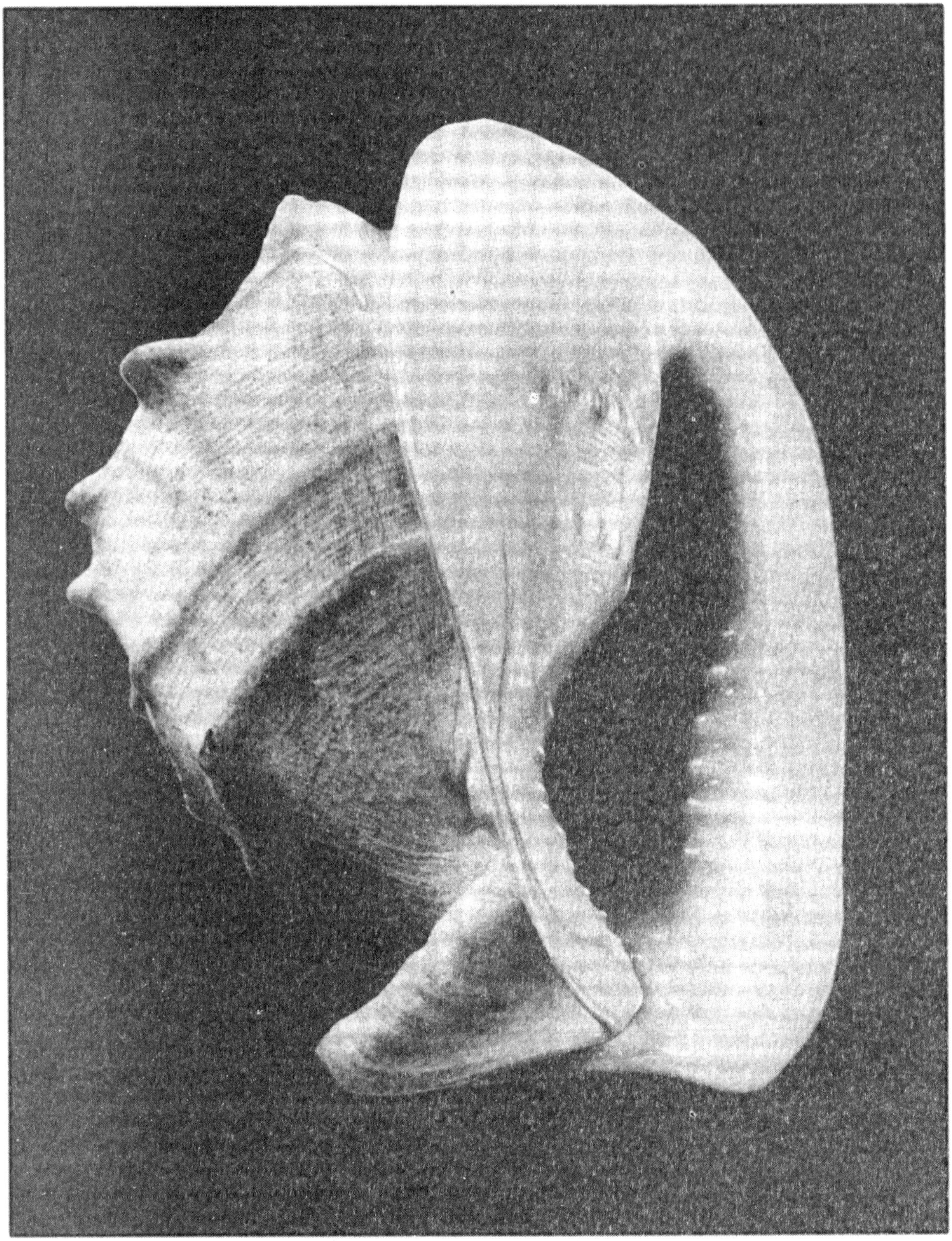

GLORIOUS DESIGN OF THE SHELL BELONGING TO A SEA SNAIL

Sea snails of the family Cassididae possess thick, massive shells, which have found much favour among carvers of cameos. The particular species seen above goes by the name of *Cassis Cornuta* and it is put on its side so that the magnificently serrated inner lips can be seen to full advantage. The spirals of the shell, seen behind, are equipped with a number of jagged projections. The rich variety and glorious workmanship found in sea shells used to be shown off under glass cases in many a wealthy home of the early nineteenth century.

Chapter LXXIX

The Wonder of the Sea-Shells

By Edward Step

Author of "Shell Life"

MUCH of the effete matter from the decay of living things on land finds its way in time to the ocean, via Nature's mighty drainage system; the sluggish brooks emptying into the swifter streams, and these into the larger rivers that flow into the sea, charged with great volumes of solid matter in solution—mineral, vegetable, and animal.

Triturated or, at least, broken fine, in the sea it finds multitudes of creatures waiting to avail themselves of this abundant food: minute transparent forms absorbing microscopic particles, and these in turn feeding larger animals, including great fishes and the greater sea-mammals, the porpoises and huge whales. Among the intermediate animals thus fed is the wonderful race of molluscs, with its infinite variety of forms that have fitted themselves for filling innumerable situations in the sea, and have modified their shapes and their protective armour to fit their mode of life.

When Man came and some of his tribes settled along the coasts they found by experiment that many of these molluscs were good eating, and were more easily obtainable than the fishes of deep water, which appear to have been left for a much later generation. In time these marine molluscs became known as shell-fish, a name shared by crabs and lobsters but no longer applicable, as the recognized fishes of to-day have internal skeletons. That early man availed himself largely of the molluscs as food is proved by the enormous percentage of sea-shells found in the ancient kitchen-middens in various parts of Europe, and by the continuance of the practice on some of the shores of Britain to-day.

A few generations back it was the correct thing for the wealthy to possess a cabinet of shells, not because the owner had any interest in natural history, but because the specimens were beautiful, quaint or rare—and the rarest were the most desirable, whether they had beauty or not. They were "curios." Mostly exotic, some that came only rarely from far eastern waters realized high prices in the sale-room. Though much admiration might be evoked by their strange forms and wonderful coloration, their importance to the animals that built them for their protection was scarcely considered; and the animals themselves did not come into the picture. The owner prided himself upon his perfect examples of the rare glory of the sea and the precious wentletrap, his big harps, helmets, mitres, cowries and cones.

Yet these same shells were marvellous in a way that their owners never realized—the method by which they had been fashioned, the purpose they served in the life of the artificer, and the wonderful mechanism for the security of those that were in two parts.

The living animal that produces these wonderful shells is always very soft; yet it has various internal organs—such as heart, stomach, liver, nervous system, and so on—analogous to those of the backboned animals. The body is wrapped by a sort of cloak (known as the mantle), which has the power of secreting carbonate of lime mixed with animal matter on its outer surface. This hardens until it is more or less stony, and successive layers are added within and at the edges as the animal's increasing bulk requires. The molluscs, as a phylum, or division of animal life, are separated by naturalists into five classes, of which we are chiefly concerned here with two classes—the snail-like gastropods and the mussel-like bivalves. It will be seen that, apart from the possession of a hard shell, there is no external resemblance

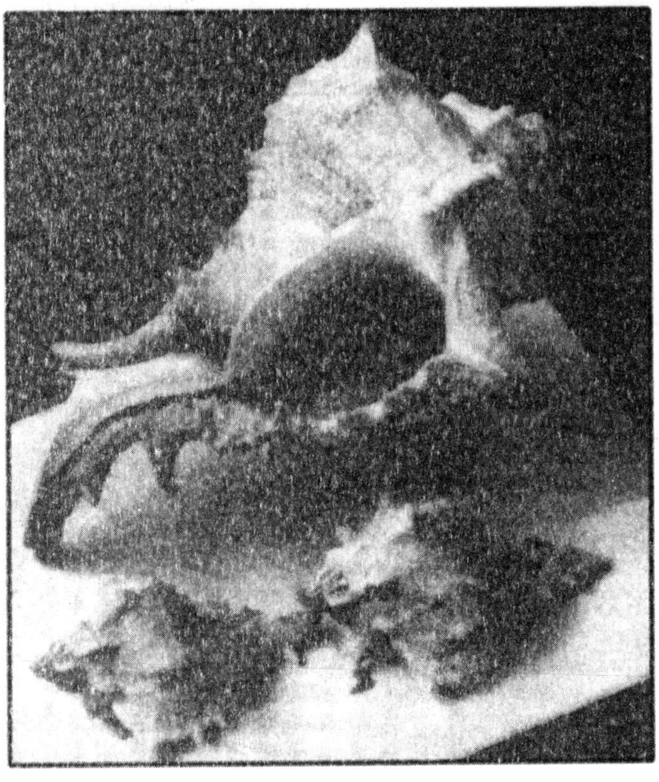

PREDATORY SNAILS OF THE INDIAN OCEAN

In the type of sea-snail called Murex the shell is commonly long and inclined to be jagged. This specimen comes from Ceylon and, like its relatives, is equipped with a rasplike tongue for boring through the shells of other molluscs on which this snail preys.

FEMALE ARGONAUT AND HER ELEGANT SHELL

Here we have a sea creature the female of which makes for herself, by means of a special secretion, a very elegant shell. In this the eggs are developed The body, here seen separated from its shell, is not unlike that of an octopus, the two creatures both belonging to the class of cephalopods. The body is from a model in the Natural History Museum. Kensington.

between these two classes; and to appearance there is little in the animals themselves.

The gastropods have the principal part of the body elongated: it is known as the foot, because its underside is flat and sole-like. At the front end of the foot is a distinct head, with eyes and other sense organs, and a mouth furnished with a ribbon-shaped "tongue" which is beset with rows of flinty hooks for rasping food. Upon the flat foot the mollusc glides over rocks and seaweeds in search of food. These animals are usually unsymmetrical, the body being coiled spirally, and the shell, being moulded upon it, takes the same form. In some species there is attached to the hinder part of the foot a shelly or horny plate (operculum) which, when the animal withdraws into the shell, closes the entrance. It might be thought that, as the mouth of the shell increased its diameter with the general growth, this tail-plate would soon get too small to act as a door; but it must be remembered that when closed the plate is in contact with the edge of the mantle, which lines the shell, and in that way receives additions to its circumference. Muscles attached to the shell pull the animal within.

THE animal that constructs and lives in the double, or bivalve shell is usually symmetrical, and its mantle is divided into two large lobes, which explains why the shell is formed in two separate pieces. As growth of the body takes place, the valves are enlarged by equal additions to their margins. This bivalve is a more sedentary animal than the gastropod. Although he has a foot he does not glide upon it, but uses it mostly for the purpose of burrowing into the sand or mud and pulling himself down with its aid. There are some exceptions, which are active after a fashion different from that of the gastropods; but the sedentary life has led to the degeneration of the class from a higher type. This decline is seen, among other items, in the loss of the head and its attachments. There is a mouth, of course, but it has no ribbon-tongue such as has been referred to: it can take in only the minute food that is brought to it in currents of water set up by the waving of filaments on the gills.

Like the naked molluscs of the land—the slugs—there are molluscs of the sea that have no shell, or only a small one concealed by the mantle. A slightly more advanced class—regarded from the shell point of view, but not structurally—has the upper surface protected by eight small, overlapping plates, margined by the mantle. These are the mail-shells, or chitons, which are able to roll up like woodlice for protection.

The simplest form of undivided shell is that of the limpets, one of them a very conspicuous feature of rocky coasts, where the animal is used as a handy bait for fishing, and in some places as human food. The limpet, consequently, is familiar to all; and it will be remembered that its shell is a hollow, low cone. Our wealthy collector of the past may not have admitted so vulgar and obscure a shell to his cabinet; and the ordinary wanderer along the shore to-day passes it by as uninteresting, though it studs the reef in thousands. A little time given to its consideration may help us to an understanding of the part that form and colour of the shell play in the life of our molluscs.

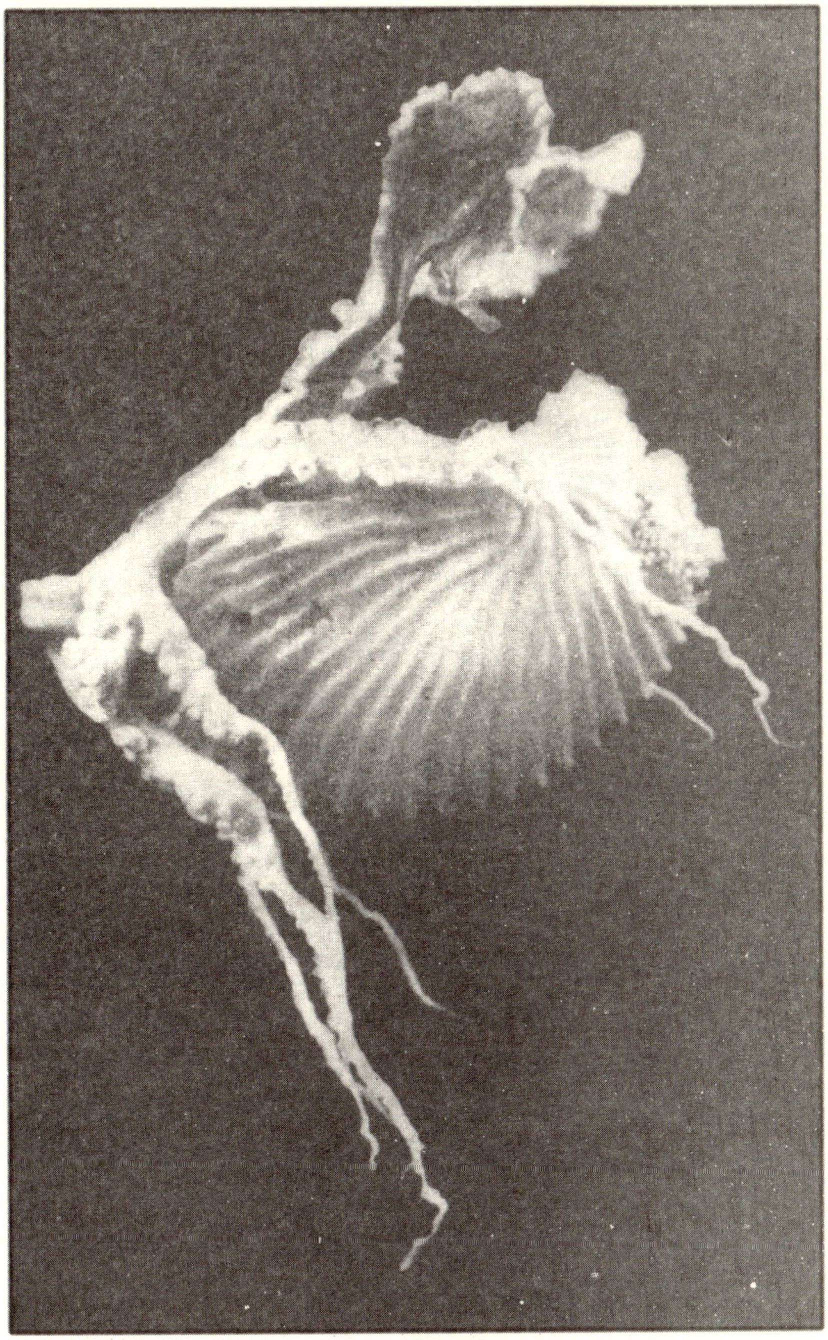

ARGONAUT OF THE MEDITERRANEAN, SWIMMING

Amid an intricate trailing of languid tentacles the little argonaut writhes gently through the sun-warmed sea. Progress is made by the expulsion of water through a special funnel, just as in the case of the cuttle-fish, a larger relative of the argonaut. The shell is formed of a secretion from two of the tentacles. This specimen comes from the Mediterranean.

If the tide has just receded from the rocks, we may see the limpet in motion, the shell oscillating a little as the animal is adjusting itself to a correct position in a depression clear of acorn-barnacles and corresponding to the outline of the limpet's shell. During the period of high-water the limpet has been wandering around with its toothed ribbon scraping off the delicate growth of young seaweeds within a radius of several feet. We can see where it has been at work by the cleared zigzag lines it has left. Now, having settled itself into the proper position, it takes firm hold with its broad, oval foot, and draws the edges of the shell down into close contact with the rock. There it will remain until the tide is running up the shore again.

Looking at the limpet in repose, we can understand that should a heavy sea spring up and attack the rocks in great force, the form of the shell will offer little resistance: the low angle will permit the waters to glide harmlessly over it. Another important feature is that in its resting position the shore-birds cannot attack it: it is only when the limpet has its shell raised in movement—mainly performed under water—that even the oyster-catcher can prise the shell from its hold and nip out the mollusc. From this tent-like form most of the shells of gastropods have probably been evolved. Several molluscs structurally different have low,

Martin Duncan

SHELLS COMMON AND UNCOMMON DISCOVERED ON BEACH AND SEA-FLOOR

As numbered here the shells in the top photograph are: 1. Cowrie. 2. Olive shell. 3. Pearl oyster. 4. Spondylus. 5. Leopard cowrie. 6. Conus gloria maris. 7. Conus acuminatus. In the lower photograph we have: 8. Razor shell. 9. Donax variegatus. 10. Venus verrucosa. 11. Venus casina. 12. Prickly cockle. 13. Queen scallop. 14. Tellina tenuis. 15. Cap-of-Liberty. 16. Otter's shell. The venus shells, of which we have two examples in the lower photograph, are found usually between high and low watermarks.

Martin Duncan

MOLLUSC HOMES OF WONDERFULLY VARIED SHAPE FOUND IN THE OCEANS

These shells (top) are examples of : 1. The venus shell—Venus casina. 2. Great piddock. 3. Prickly cockle. 4. Lima nians. 5. Ormer or ear shell (showing the outer surface). 6. Short piddock. 7. Otter's shell. 8. Cap-of-Liberty or heart cockle. Below are : 9. Sea snail. 10. Winkle. 11. Tusk shell. 12. Boat shell. 13. Pelican's foot shell. 14. Top-shell. 15. Ormer or ear shell (inner surface). 16. Great whelk. 17. Ladder shell. 18. British cowrie. 19. Dog whelk. 20. Screw shell. 21. Rock winkle. 22. Limpet.

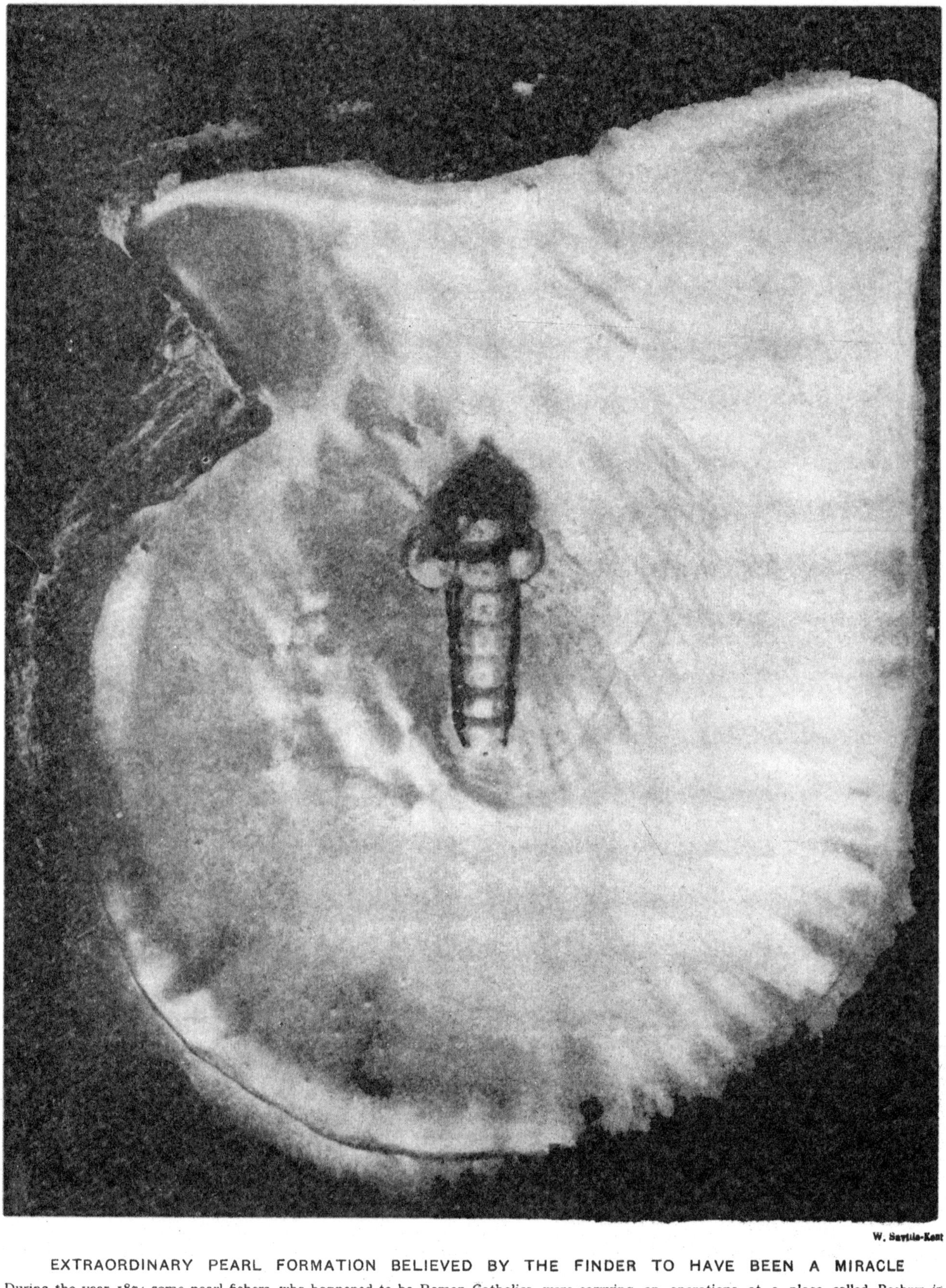

W. Saville-Kent

EXTRAORDINARY PEARL FORMATION BELIEVED BY THE FINDER TO HAVE BEEN A MIRACLE

During the year 1874 some pearl fishers, who happened to be Roman Catholics, were carrying on operations at a place called Raeburn in Western Australia. One of the number came upon the shell we see above and, on opening it, saw what he believed to be a miraculous sign. By some freak the oyster had formed not one but nine pearls, and they were arranged in the form of a cross. These joined pearls received the name of "The Southern Cross," and became famous in the annals of the industry.

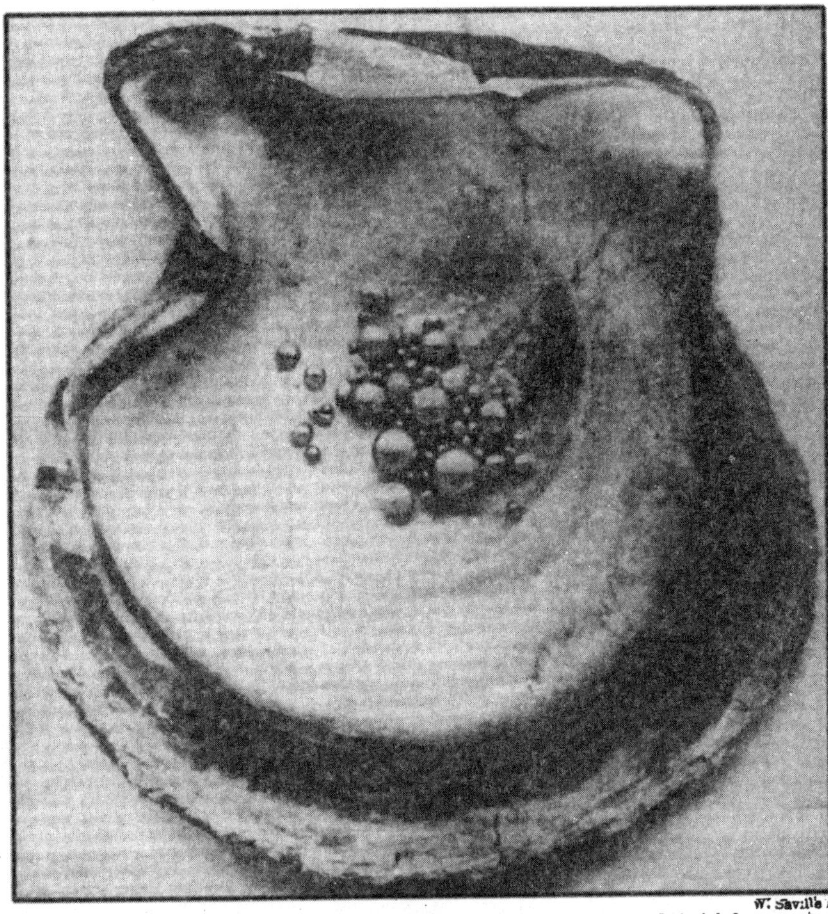

W. Saville Kent

AUSTRALIAN PEARLS AS FOUND IN THE SHELLS

These pearls come from Shark's Bay, a deep depression in the coast of Western Australia largely protected by islands from the force of the ocean. Here there has been great activity in the pearl industry. The pearl itself consists of layers of calcium carbonate laid down either round some foreign body, such as a grain of sand, or round the body of a parasite.

smooth shells that conform to the same principle of non-resistance. Where more room was required for the safe housing of the resting animal the low cone became a comparatively high cone; but where similar exposure on shallow-water rocks would make this dangerous the cone was coiled obliquely and the general form was retained. This may be seen in the common top-shells and periwinkles. Some of these show, also, modifications of their sculpturing to fit their differing modes of life. Those that are, from their usual situation, most exposed to wave action have very low spires, and the depressed cone is smooth, whilst those that frequent sandy or pebbly shores where their foothold is precarious have the shell decorated with knobs and ridges to prevent them from rolling unduly when swept by the waves. Corresponding adaptation will be found in all the common shells of the shore, whether the animals live by cropping the tender seaweeds and zoophytes from the rocks, feed upon the leathery wracks that are torn and twisted by the waves, or hunt for animal food in sheltered pools where there is very little disturbance of the waters.

Some of the limpet-like shells have an aperture at the top or a slit in the edge for the admission of new water when the shell is closed down to the rock. Some, with evidently the same ancestry, have this aerating slit tucked into the thick lip of their spiral shells. In the whelks and many others of deeper water habit, the cone is very elongated, owing to the necessity of making constant additions to the newest part to accommodate the rapid growth of the animal. The upper part of the spiral may be quite unoccupied; and in some species the empty room is shut off by a partition being formed across it. Where less space is required in what is known as the body-whorl, the spire may become almost cylindrical and of great length relatively to its diameter, as seen in the British auger-shell, horn-shells, and needle-whelk.

The pelican's-foot shell in its younger state approaches this slender form, but when older the margin of the lip spreads out into several broad lobes, covering all but the easily-retracted head of the animal, which is extended into a proboscis. The tooth-ribbon of this and other carnivorous species is used as a drill which bores neatly through the shells of limpets and others upon whose inmates the borer feeds. The dog-whelk has similar tastes and methods; and the sting-winkle, whose shell is all rugged knobs and folds, getting a living by like means, is one of the greatest pests to the oyster-farmer on those parts of the coast where it is very common.

The alteration in form effected in the shell of the pelican's-foot between youth and maturity is not as great as that attained by a mere shifting of the opening of the shell more fully to the side, as may be seen in our little margin-shell, where the body-whorl accounts for about four-fifths of the entire length of the shell, the remainder being a small squat spire. The mouth is a narrow, undulating opening at the right-hand side of the body whorl, the lip at first thin edged, but later much thickened and toothed. Now, if this lip were much expanded outwardly it would have the effect of bringing the mouth-slit nearly into the centre: and that is just what we find has happened to the shell of the cowries, of which there is a small but pretty representative on British shores. In the warmer oceans there are large cowries with fine coloration and a natural

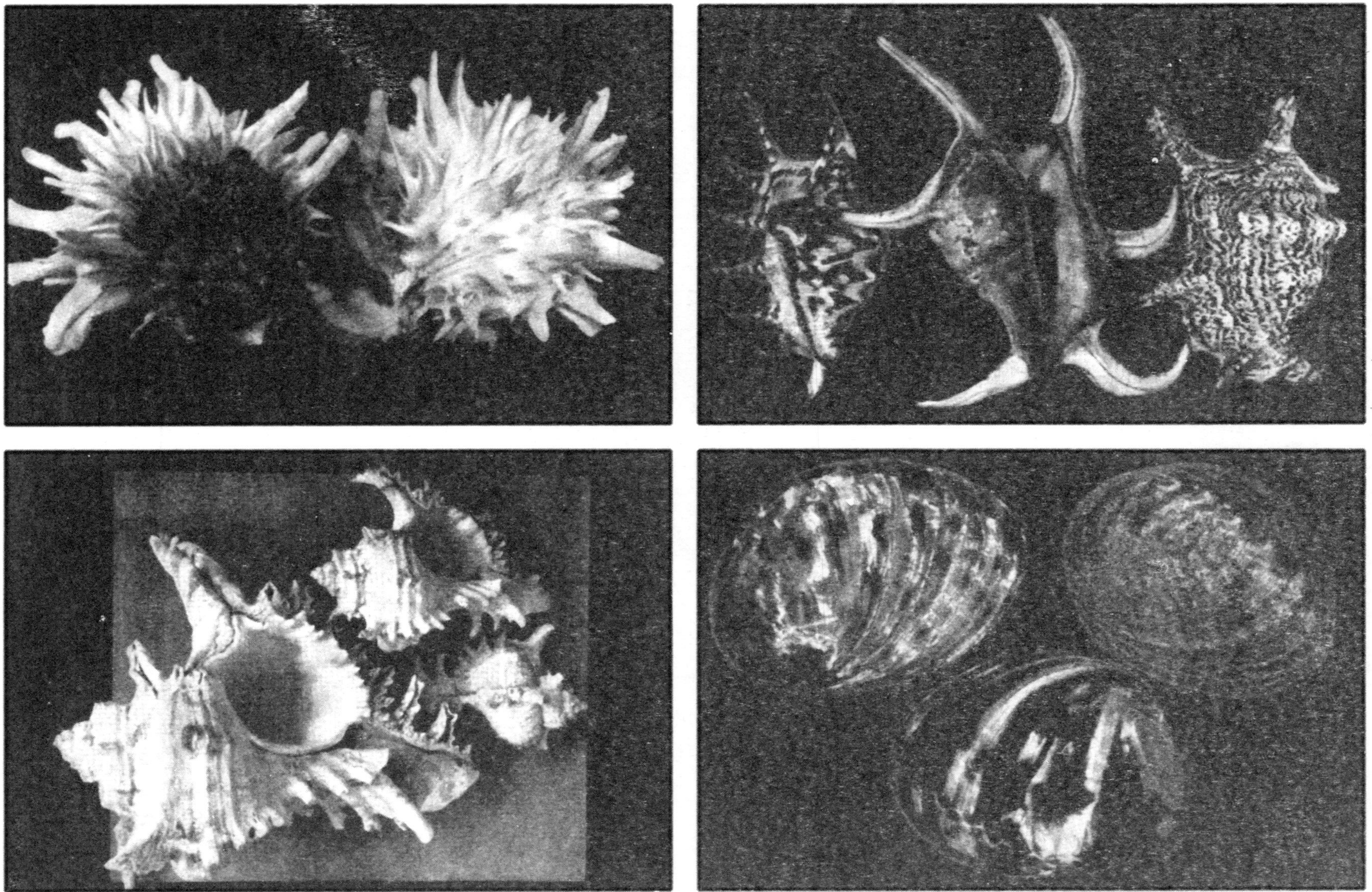

J. J. Ward

THE WEIRDLY WROUGHT AND STRANGELY TWISTED HOMES OF SOME MOLLUSCS OF THE SEA

In this quartette of shell photographs we have (bottom left) a specimen of Ramosus from West Africa and (bottom right) some ormer or ear shells. They are sometimes washed up on the south coast of England and, in the Channel Islands, the animals inside are used for food. There are a number of perforations in the shell through which tentacles protrude when the creature is alive. These holes are the remains of what, at one stage in this mollusc's history, was a continuous slit in the last whorl of the shell. There is a corresponding slit in the mantle, that is the thin fleshy membrane which lines the shell. The shell of the thorny oyster, spondylus (top left), is beautifully coloured. The top right-hand illustration is of some wing-snails.

W. Saville Kent

SOBER COLOURED OUTSIDE OF A MOTHER-OF-PEARL SHELL

The exterior face of any shell in which mother-of-pearl is found does not hint at all at the glories within. As we see here, the shell seems to be composed of a number of strata, and there is a series of projections arranged in vertical rows. All this rough outside has to be scraped away until, gradually, the wonderful mother-of-pearl is arrived at. This has much the same composition as pearls. In the East they carve these shells—the finest specimens—with intricate figures so that the light, seen through the shell, varies with the depth of the carving.

high polish, many of them treasured as parlour ornaments when brought home by travellers. A smaller, yellowish cowry from the Pacific is known as the money cowry from the fact that traders in Africa used it extensively as a medium for bartering.

OTHER gastropods with long expanded mouths keep them to the side, as in the large but light tun-shells and the heavier volutes, or music-shells whose internal coils act as a whispering gallery, giving out faint sounds which are fabled to be the murmur of the distant sea—Wordsworth's

> Murmurings, whereby the monitor expressed
> Mysterious union with its native sea.

As a contrast to these thick, heavy shells we have the keel-shell, a semi-transparent, open shell, though still spiral in its structure. It looks like a delicate piece of thin blown glass, so fine and rare that collectors of a hundred years ago were willing to give a hundred pounds in exchange for a specimen no bigger than half a walnut-shell, over which one must breathe lightly for fear it might be blown away. Its rarity and its delicacy are explained by the habit of its maker, a curious elongated creature with its foot developed into a fin, which enables it to swim in the upper waters and feed upon the smaller jelly-fishes and other floating animals, with the exquisite shell below it. Even to-day, when the habits and habitats of the animal are better known, a purchaser has to pay several pounds for the wrinkled shell.

Shells as delicate but more plentiful, and therefore of less value commercially, may be found on British shores; such are the Marsenia, the lobe-shells and bubble-shells; but they must be sought when the animal is alive. They are then enfolded by the mantle of the mollusc, which feeds upon the jelly-

S. C Johnson

J. J. Ward

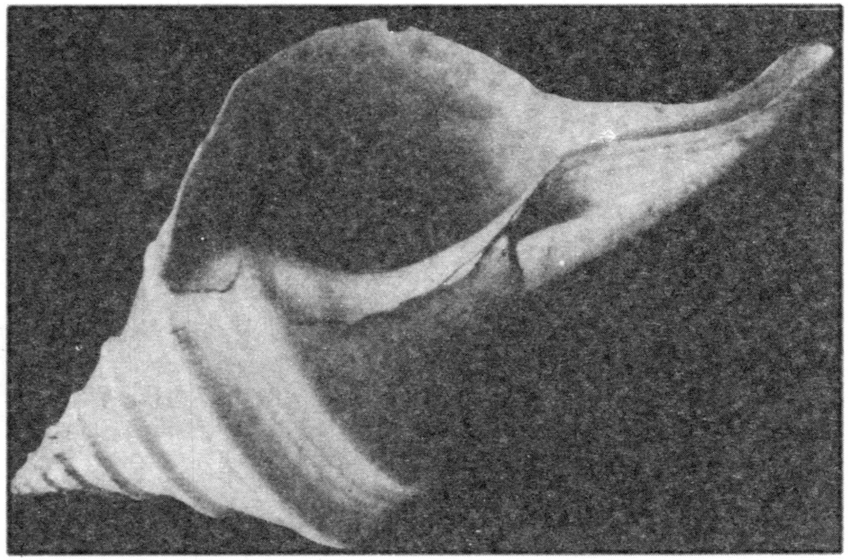

LOVELY SPECIMENS OF NATURE'S ART THAT ARE FASHIONED UNDER THE WAVES

The shells seen here are (bottom left) Spondylus Americanus; (bottom right) Fusus Proboscidiferus, which has a very elongated mouth and a spiral built up like a pagoda; an ormer shell (top left) with its row of perforations like port-holes in a ship's side and (top right) the shell of a paper nautilus. This last is one of the wonders of the deep and has a translucent shell, beautifully ribbed, which serves to hold the eggs. The shell is not attached in any way to the creature it serves. Another name for the paper nautilus is argonaut, though strictly speaking it is only the Mediterranean species to which the name paper nautilus should be applied. The argonauts are members of the cuttle-fish family.

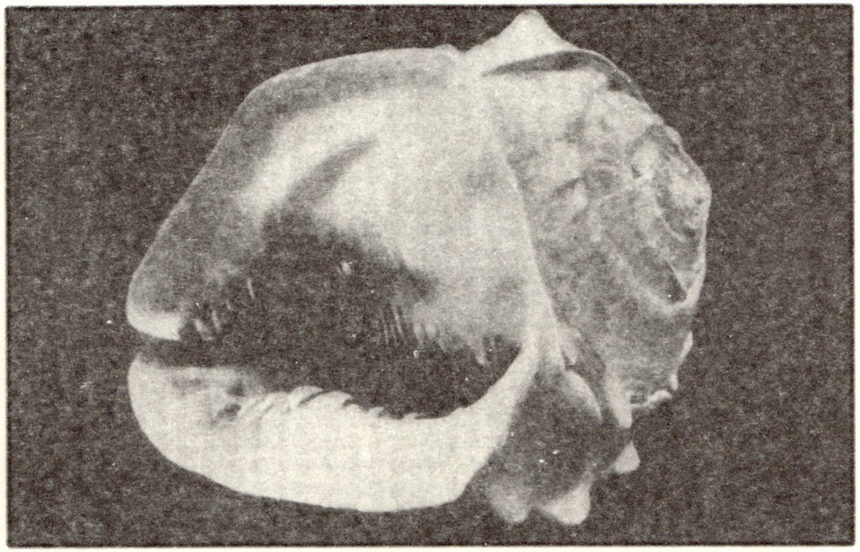

DIVERSE AND BEAUTIFUL SUITS OF ARMOUR FOR LOWLY CREATURES OF THE SEA FLOOR

In this page we have a model (bottom left) of Murex Ramosus from the Natural History Museum at South Kensington. The creature is inside and its protruding " stalks " and the " foot " on which it moves are all the living portions visible. The lower right-hand photograph is of a specimen of Triton Tritonis with an elegantly fluted mouth and a finely tapering whorl. Above are (top left) a specimen of Sinensis, a shell found in the seas between China and Japan, and (top right) the wide-mouthed Cassis Cameo from the West Indies. The word " cameo " in the nomenclature of the last-named shell reminds us that the exquisite work of the cameo artist depended for its perfection on the quality and texture of certain shells.

Papuina shell from New Guinea

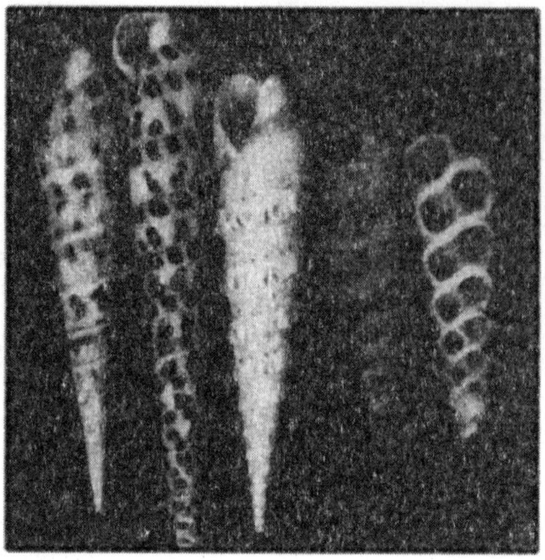

Auger shells

Deep-sea Pacific shell

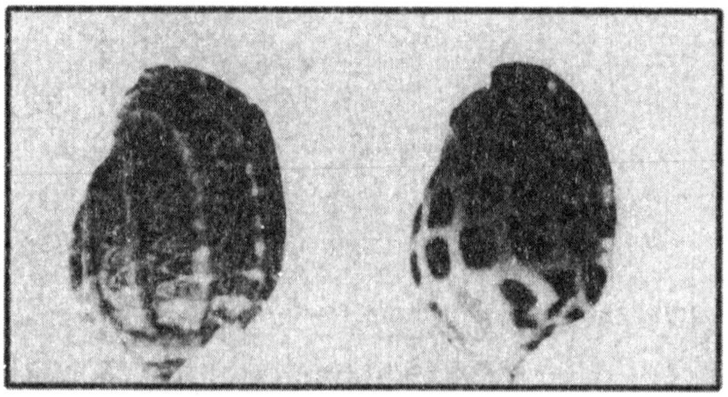

Cassidaria (left) and Cypraea (right

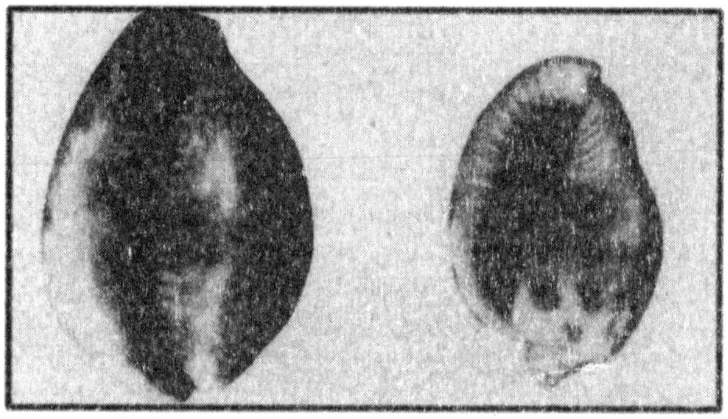

Cypraea (left) and harp-shell (right)

Volva shell

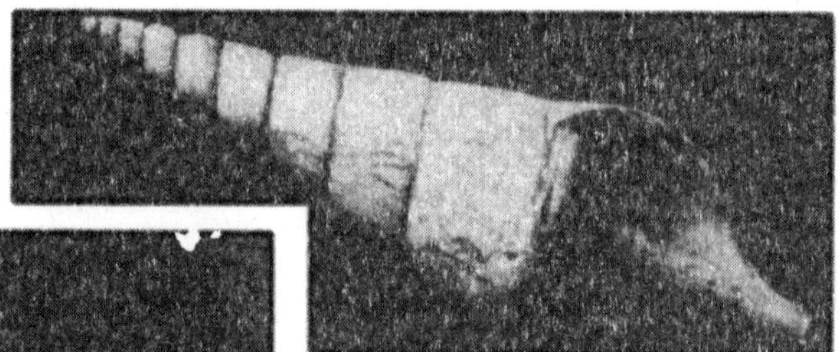

Rostellaria

Pheasant shell

Scallop

Polynesian cone-shell

SEA MOLLUSCS IN THEIR INFINITE VARIETY THAT ARE FED ULTIMATELY BY THE LAND

Life goes on under the sea much more intensively than on dry land. The oceans teem with life, but it must be remembered that this life draws much sustenance from the land. Water vapour, sucked up from the sea by the sun, discharges itself as rain upon the earth, and flows back to sea again in the rivers. In doing so it brings down with it a constant supply of decaying particles of what were living things, and this feeds a host of minute forms and these in turn feed larger creatures. Among the intermediate forms thus fed are the sea-molluscs.

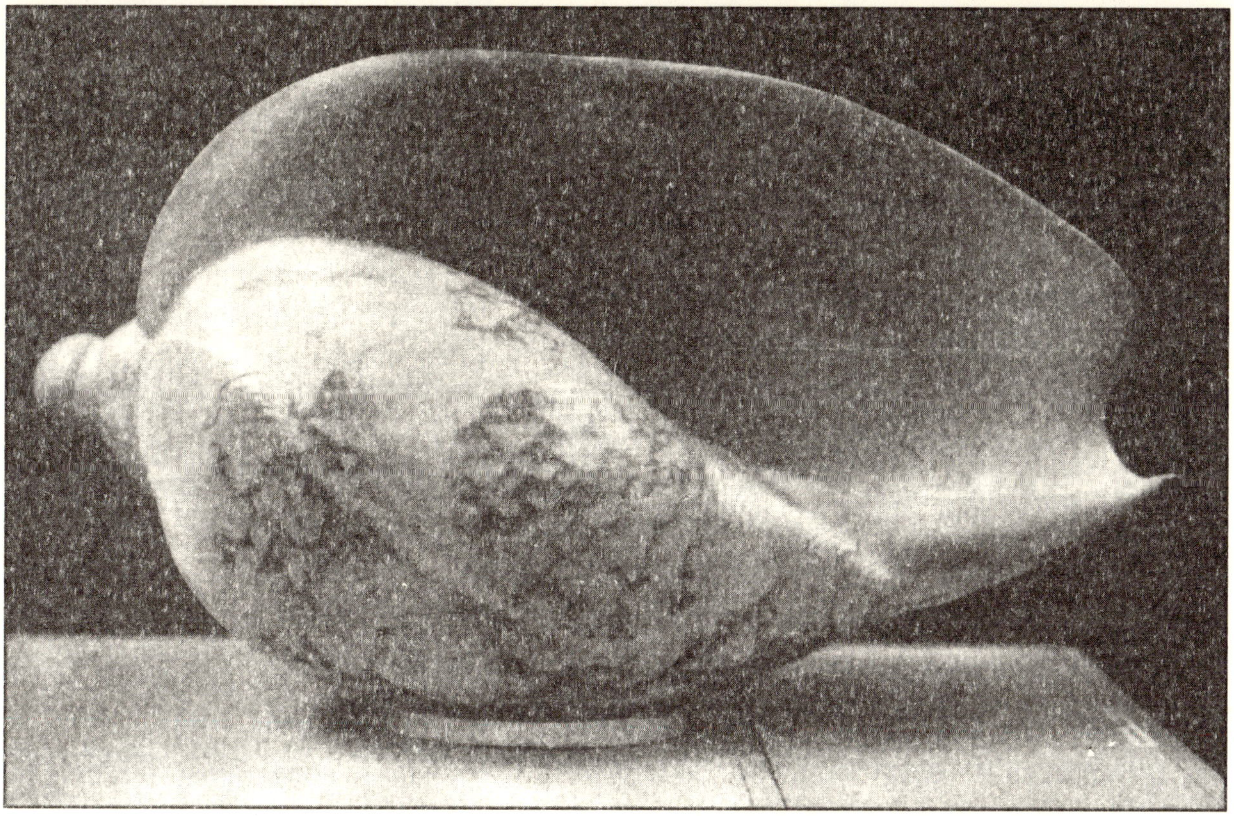

SHELLS FROM TASMANIA AND HAWAII, WITH THEIR BEAUTIFUL MARKINGS

In the lower photograph we have an example of the Volutidae, which is the name for members of a division of the mollusc kingdom including many most elegant shells not only handsome in shape but also in colour and markings. This particular form, *Mamilla*, comes from Tasmania. The upper photograph is of a species of tun-snail. A feature of this class of molluscs is that the glands connected with the mouth secrete a liquid containing sulphuric acid which aids in the process of perforating the shells of other molluscs.

E.N.A.

VARIATIONS IN THE BEAUTIFUL TREE SHELLS

These little creatures live on the leaves and the bark of trees growing in the remarkable Rain Forest which clothes the mountains overlooking Honolulu. Many of the shells are most vividly tinted, some displaying pink or green, while others are tricked out in red, gold and brown Such brilliance of colouring is more usually associated with sea than land molluscs.

It is well known by every reader that a large class of molluscs have the shell formed in two pieces, or valves, and are known consequently as bivalves. Of these the most familiar are the oyster, scallop, mussel and cockle. What has been said of the prevailing symmetry of bivalves scarcely fits the case of the oyster, which has unequal valves, the left, or lower valve being hollowed out and holding the animal, whilst the right, or upper valve is nearly flat. But the oyster in its infancy attaches its lower valve to the rock or stony bottom, and remains fixed for life—or until removed for the food of man.

In some species the two valves are kept closed tightly by the action of a strong central muscle that unites them, as in the oyster and the scallop; but in most species there are two such adductor muscles, one towards each end of the shell. The valves are hinged together by a band of soft horny material which, but for the pulling action of the adductors, would cause them to gape apart. To give greater stability and security to the closed valves, there is just below the horny hinge a more or less intricate series of ridges and depressions in the solid material of one valve which fit with wonderful accuracy into corresponding ridges and depressions on the other valve.

Many of these bivalves live buried in sand or mud, into which they are pulled by the action of the muscular foot; some, however, attach themselves to rocks by tough threads (byssus) spun by the foot. A familiar example is a crowded cluster of mussels on the wooden piles of a seaside pier.

like ascidians coating the submerged rocks. Similarly delicate but much smaller shells are those of the butterflies of the sea, so called, quaint molluscs that float on the surface and feed upon the minute life (plankton) there so abundant, though mostly invisible to our eyes. These sea-butterflies (*Pteropoda*), in their turn, form a considerable part of the food which nourishes some of the very largest whales.

Great beauty is displayed by many of these bivalves, either in sculpture or colouring, or both in the same individual, according to the demands of their life. Others, like the big pearl-oyster, which converts intrusive rubbish into precious gems, have their shells richly ornamented within, but may be commonplace or even squalid-looking outside.

Chapter LXXX

Animal Hands and Feet

How the Animals got them and What They did with Them

By T. H. Gillespie

Director of the Zoological Park, Edinburgh

To find the origin of feet and hands one must look back for a good many million years to a time when a new kind of animal had just come into existence. It was designed on a new plan, and it had a continuous chord of nerve matter extending the length of its body and the rudiments of a skull and brain, and it was the highest form life had yet attained—a primitive fish form.

Those earliest fishes were of simple elongate form with little about them to suggest the great destiny awaiting their descendants, and their eel-like progress through the water was aided by an expansion of skin running along the back and round under the tail. Soon, in some forms, this continuous middle fin was, it seems probable, supplemented by a similar expansion of skin on each side to help to keep the body upright in the water, something after the fashion of the bilge keels in the hulls of ships.

As need for increased power and speed was felt it appears to have been advantageous that these continuous fins should be broken up, parts of them being suppressed and the fragments remaining becoming larger and stronger and being reinforced by numerous rays, or rods of cartilage. Of the central fin several portions were retained, the most important being the tail fin, the fish's propeller. Of the lateral, or side fins only two fragments were retained on each side, forming two pairs of paddle-like fins. These were Nature's first rough sketch of what were eventually to become hands and feet.

W. S. Berridge

APE WHOSE FEET ARE HANDS

Look at this orang utan clinging to a sapling. The feet are supporting it below as efficiently as its hands are holding it above. One hand is visible with its back and knuckles towards the camera while the other is above the animal's head, and out of the photograph.

Doubtless these paired fins were very useful to their owners in balancing and steering, which were probably the needs that stimulated their production, though one might imagine that the swimming fish could have got along without them. Their greater importance began when some enterprising and adventurous fishes, pursuing their prey in the shallows, ventured to continue to hunt on to the wet mud or sand. Then the two pairs of side fins, pressing on the mud, became indispensable for maintaining an "even keel," and as more and more stress was laid on them they became gradually stronger and better able to support the body and assist its flopping mode of travel. To this day there are fish such as, for example, the mud-skippers and climbing perch, which habitually leave the water and use their strong pectoral, or breast fins in travelling over mud flats or climbing rocks or even trees, and we may see in them, perhaps, some suggestion of the manner in which the ancestors of the higher animals—reptiles, birds, mammals—first stepped ashore.

There was, however, a very long road to travel from such fins

H. Bastin

FEET AS EVOLVED BY THE BEETLE AND THE CRICKET
Here is a horned chafer beetle from East Africa (bottom) with its leg stretched out. Above are enlarged photographs of (left) the claw or foot of a goliath beetle and (right) the fore-leg of a mole-cricket. The toothed "foot" works with the next joint like scissors to cut a way through the roots of turf in which the insect burrows.

to the limbs of even the most primitive four-footed land animal. Gradually, as more and more strain was laid on the paired fins, they would respond with increasing strength; the rods supporting them would, by fusion one with another, become stronger and harder as they became fewer in number; then the need for flexibility in the thickening and hardening rods would lead to the formation of articulated joints. It is probable that even before the ascending types had passed from the fish stage the paired fins had acquired something of the form of limbs, with a jointed shaft connected to the body skeleton by incipient shoulder (pectoral) and thigh (pelvic) girdles, and having at their lower ends the fin rays, much reduced in number, shortened, strengthened, and hinged or jointed.

As the earlier fish-amphibian forms came into being the number of fin rays or, as we ought now to begin to call them, digits or toes, were reduced to, perhaps, eight, then to seven or six, and ultimately, by the time the amphibians (represented to-day by salamanders, frogs, and so on) were definitely established, their toes or digits had been brought to what was evidently to be a standardised number of five on each foot. Yet to this day human beings are sometimes born with six toes or fingers, showing, perhaps, the persistence of the age-old inheritance.

Nature, however, has the itch of experiment and, having produced this standard five-toed foot, immediately began to modify and alter it. In a general way one might say that there was underlying much of the ensuing alterations the idea that four feet and twenty toes were superabundant for locomotion alone and that some of them might be put to better use. Among the amphibians such modifications as occurred were not particularly exciting or interesting; they took the general line of mere suppression of some of the digits or even of the entire limbs. Some of these creatures, indeed, reverted to an entirely aquatic life and re-acquired the long body, with flattened tail and expansions of skin for swimming, with a strong resemblance in form to their fish ancestors.

NATURE, having brought her new backboned animals ashore by the medium of the amphibians, seems to have regarded that class as of little further account, and from some early salamander-like amphibian she very soon produced another new kind of animal—the reptiles. It was when the reptiles had become established that more important changes in the form and use of the limbs began to appear. The fashion, so far, had been to use all four feet for

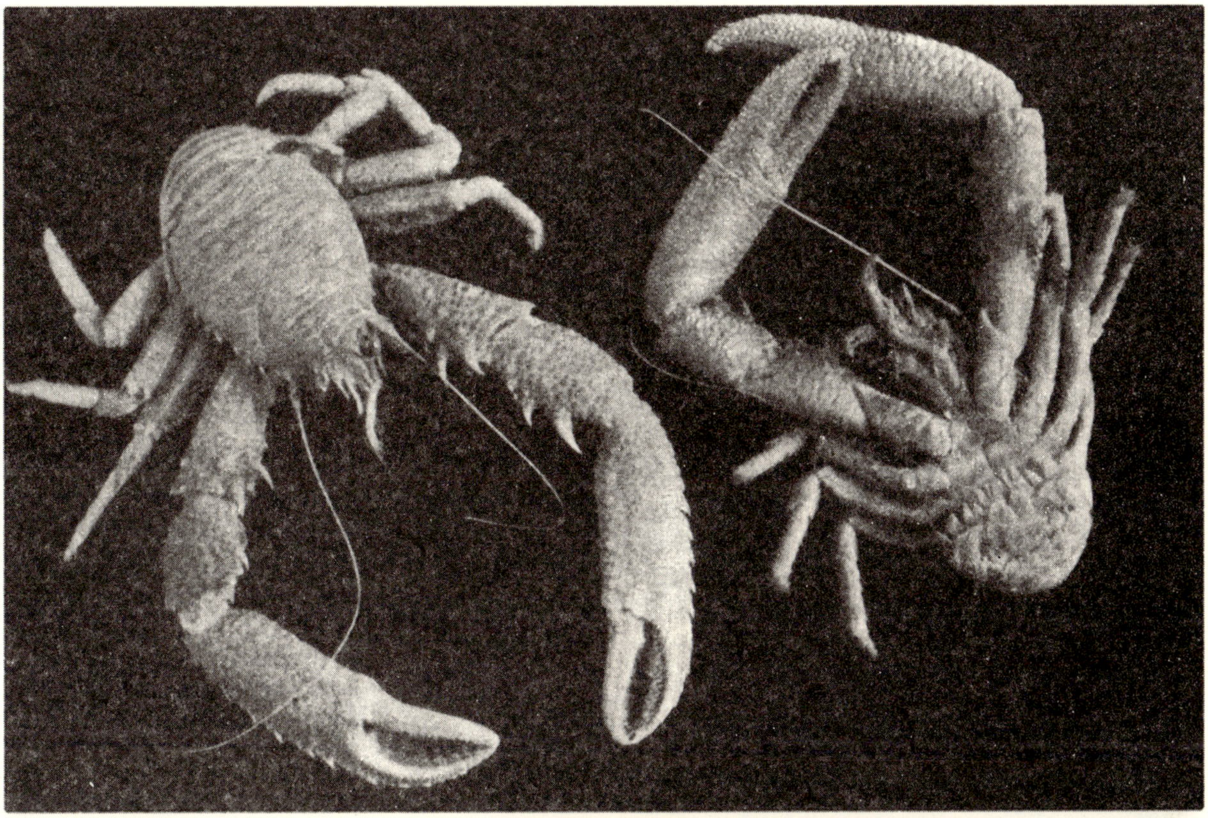

Martin Duncan

walking, though even among the amphibians the forefeet may have been used to assist in getting food into the mouth, as one may see in the case of present-day newts and frogs.

Some of the old reptiles went much further than this, for they learnt to sit up on their hind legs and hold their food with their forefeet. As the use of their forefeet in this way developed and more of the burden of carrying the body was thrown on the hind legs, the latter became relatively longer and stronger until, in some groups, the forelimbs probably ceased to be of any but the slightest use for locomotion, and these creatures walked or hopped on their hind legs. One has only to look at a squirrel nibbling a nut or a kangaroo crossing a paddock to realize something of the appearance of those old reptiles and the way in which they used their front feet. That was a quite definite attempt to make a real hand, but, having got so far, it never attained complete success and the reptile forms in which it appeared became extinct. No living reptile uses its forefeet in such a manner, though it is interesting to notice that several species of lizards which, in the

STRONGLY GRIPPING CLAWS OF CRAB AND LOBSTER

In the lower photograph a land crab in the London Zoo is demonstrating its gripping power by supporting its entire weight with one limb. The top photograph displays two squat lobsters with ugly-looking barbed claws. On the left we have a view from above and on the right we see the underside.

Shepstone

PRIMITIVE FORMS OF HAND AND FOOT: A CLUSTER OF BOWFINS

Clearly seen through the glass plate of their tank in the aquarium this cluster of moving fish which swim under and over each other in a welter of tails and fins and lithe, wet bodies displays to us the early beginnings of the hand and the foot. The origin of hands and feet goes back to the time of an eel-like sea creature which had the rudiments of a skull and brain. Its movement through the water was aided by an expansion of skin running along the back. The bowfin still has a long fin something like the one possessed by that ancestor

ordinary case run on all four feet, will at times, especially when travelling for some distance over level ground, rise on their hind legs and run on two feet, the weight of the body being counterbalanced by the long and heavy tail.

In another branch of the reptiles was tried the experiment of adapting the forelimbs for use in an entirely new and very important direction—they turned their hands into wings and acquired the power of true flight. The wings of these flying reptiles were formed of a skin, or membrane, and to outstretch and support it the fourth finger of each forelimb became enormously lengthened, and to the group was given the name "pterodactyl" (wing-fingered). The skin, or membrane was continued to the hind legs and tail, but the toes of the hind feet and the remaining fingers of the forelimbs remained free and capable of independent action. This was a great achievement, but even with the advantages that flight affords, these flying reptiles failed to hold their place and gradually

Shepstone

Schensky

HORSE MACKEREL AND RAINBOW TROUT AND THEIR FORMS OF FIN

While the hand and foot began their course of evolution in the sea and ended on the land, certain forms of life, the fish, retained these primitive forms of hands and feet, different species developing their fins in different ways. Compare the fins of these horse mackerel (bottom) which live in the sea, to those of the fresh-water rainbow trout (top). The lower photograph was taken by F. Schensky and is from "Animal and Plant Life in the North Sea," which is published by the Biological Station at Heligoland.

Otho Webb

W. S. Berridge

HANDS AND FEET AS DEVELOPED IN THE GREEN AND SMOOTH-CLAWED FROGS

As the earlier kinds of fish-amphibian forms came into existence during the course of evolution, the number of rays on the fins, which were the beginnings of what in later animals we call digits, or fingers and toes, was reduced to eight or six. By the time that the true amphibians, represented to-day by the frogs, had been evolved, the number of digits had become standardised by Nature to five to each foot or hand. Here are two examples of the primitive hand in the smooth-clawed (bottom) and green (top) frogs.

W. S. Pitt

CLAWED HANDS AND FEET OF THE GEOGRAPHIC TERRAPIN OF NORTH AMERICA

Geographic terrapins are natives of North America and live in fresh water. Here we see another type of hand with the digits developed but with no power of gripping in them. Their use is for swimming and crawling. The geographic terrapin is remarkable for the variations in the marking of the under-shell, no two specimens being quite alike in this respect. The attitude of the one in the lower photograph, which was taken under water, is rather reminiscent of a " levitation " as pictured by one of the old religious painters.

W. S Berridge

CLAWED TOES OF THE HAWKSBILL TURTLE AND GALAPAGOS TORTOISE

With the flipper of the hawksbill turtle we have a kind of half-way-house on the road of development from fin to leg and foot. The "feet," if they may be so described, merge imperceptibly into the "legs." The hawksbill turtle is the chief source of tortoiseshell, but useless as food. In the Galapagos tortoise (top) we see strongly-clawed feet and the legs still retaining the scales inherited from the early finned animals of the primeval seas. Indeed, these limbs seem better adapted for swimming than walking.

died out. That two such promising adventures should end in failure may have been due, in part, to limitations of the reptilian body and brain.

So far all animals had been "cold-blooded," that is to say, they were unable to sustain a temperature of body above that of the air or water surrounding them, and they had been covered with skin alone or with skin defended by scales or horny or bony plates. Long before the age of reptiles had reached its zenith, however, a new experiment of a most revolutionary kind had been proceeding and new types were now evolving with improvements in heart, lungs, and circulation of blood, bringing the capacity to maintain a constant temperature of body independent of that of its surroundings. With this capacity came greater potentialities of adaptation of body and growth of mind. To conserve this temperature something more heat-retaining than skin or scales was required to cover the body, and there appeared in due time the hair-covered mammal and the feather-covered bird.

The birds converted their forelimbs into organs of flight, but far more efficient ones than those of the flying reptile because they were driven by a better engine, and this entailed a greater change in the structure of the limbs so that, in order to give the fullest support to the flight feathers of the wing, the bones of the fingers became more or less fused together.

By turning their hands into wings of such an efficient kind the birds were made masters of a new world, the air; they could travel for distances and at a speed that nothing running on the ground could dream of and they had thus greatly enlarged powers of obtaining food and of escaping dangers. So they have held their own and have become the most varied and, perhaps, the most numerous class of animals.

Some groups of birds fell aside in the race, and either never attained complete ability to fly, or, if they did so, they afterwards lost it. Living examples of such birds are the penguins, in which the wings have been turned into paddles or flippers—wonderfully efficient for swimming but utterly useless for

W. S. Berridge

M. H. Crawford

A GREEN LIZARD AND AN INDIAN GECKO

This green lizard (bottom) comes from the Channel Islands and its general shape reminds us of those huge dinosaurs that were the limit of development, so far as size was concerned, made by the reptilian forms of life. Above is a verticellated gecko from India. Its feet are very interesting, their shape being rather like that of some cactus plant.

THE CLAWS OF THE PERCHING HERON
When perched the heron's claws are seen to the best advantage. They are very long and powerful and the nails look really formidable. Notice that the claws to the front hang a long way below the perch. This is a white heron, or egret, yielder of the famous plumes used in millinery.

flying—and the ostrich group in which the wings are quite feeble and serve little apparent purpose save that of ornament.

The hind limbs of the birds remained constant to their original duty of carrying the body, but since the forelimbs were entirely cut off from any hope of functioning as a hand we find that there is among birds a distinct tendency to use the feet for this purpose; among parrots, for example, one may see quite small objects readily held in the foot. In the feet of birds many very interesting changes and adaptations have occurred.

In the first place, the several bones which originally formed the foot have become fused together and lengthened, and so what many people regard as the bird's leg, that is the scaly part (reminiscent of its reptile ancestors) which reaches from the toes to the joint (the ankle) where the feathers usually begin, is in a sense, its foot, and the bird stands and walks on its toes. The toes are never more than four, the fifth having been lost, and in most birds they are arranged three towards the front of the leg and the fourth (which corresponds to the first, or great toe of our own feet) projects behind. In the perching birds and the birds of prey this hind toe is as strong as the others and of great importance since the grasping power of the foot depends on it, but in some groups, such as the swimming and the running birds, it is reduced in size and useless, and in some it is absent altogether.

In the perching birds the tendons which bend the toes are so arranged that when the ankle joint is bent the muscles that contract the toes are tightened and the toe drawn together. By this means, when the bird settles down on its perch the toes grip it without any muscular effort, and the more sleepy the bird is and the more heavily its body presses downwards the firmer is the grip, and so the bird sleeps in safety without fear of falling. In the eagles and falcons the gripping power of the toes is amazing, and these birds seize and hold their prey with their feet, the long, sharp claws with which the toes are armed adding to their efficiency. In these birds the first, or hind toe is of great importance, and a falcon's ability to gain his living is much reduced if any accident should deprive him of a hind claw. Then the swimming birds have altered their feet by connecting the toes with a webbing, or membrane, so that the foot acts as a paddle. Most of the swimming birds, even those that dive and swim below water, such as cormorants and gannets, have kept their wings in good order for flying, the penguins being the only group of aquatic birds that are incapable of flight, and they also have the toes webbed, though their feet are not used in swimming.

Most interesting is the change that has taken place in the foot of the ostrich. This bird, utterly incapable of flying, depends entirely on speed of foot for safety. The call for speed has led to great development in the length and strength of the legs, but it has done something more. It is clear that Nature places more reliance on one strong organ than on the cumulate strength of several weaker ones, and so we find that the ostrich, having to carry a heavy body far and fast, has reduced its toes to two on each foot, and not only so but that it seems in process of getting rid of two of those that remain, for one toe on each foot is much smaller and weaker than the other, and does little more than touch the ground, and almost the whole weight of the bird is carried on two very strong toes. A similar modification has taken place in the foot of the ostrich as that undergone by the feet of the horse, though in the bird it has not yet proceeded quite so far along the path of evolution

HANDS AND FEET AS DEVELOPED BY THE GEOGRAPHIC TERRAPIN

This brightly coloured terrapin of North America uses its hands and feet not only for swimming but also for crawling. The digits, or toes and fingers, are therefore provided with claws. Nature has been particularly busy with subtle modifications of the hands and feet of the chelonians, to which order the terrapins belong. The geographic terrapins have beautifully and very variously marked undershells.

Now we come to the mammals, the most interesting class of all since we ourselves belong to it, and it includes the widest diversity of size, form, habit, and intelligence, of locomotion and of feet; moreover, it is among the mammals that the age-long effort to produce a true hand finally reached its fulfilment.

The mammals are descended from some reptile stock of generalised type, perhaps from something very near the amphibian; at any rate, they evidently inherited the full complement of five toes to each foot. They soon began to possess the earth, and as they increased in number and competition became keener they adopted a multiplicity of ways of gaining a living with equally numerous related adaptations of form. Some of the paths they chose show a strong resemblance to those followed by the reptiles. One of the most primitive types still existing, the kangaroos, have very long and strong hind legs on which they travel by leaping instead of running, and diminutive forelegs which are used chiefly for holding food or grasping, and in these respects they recall what one knows of the old bi-pedal dinosaurs. The mammals also made a bid for a place in the air and the flying pterodactyls are paralelled by the ancient family of the bats. In turning their forefeet, or hands into wings the bats have retained the full number of fingers but excepting the thumb, which is of moderate length, the finger bones are all exceedingly long and slender, and support a delicate membrane which stretches along the arm, over the fingers, and back to the hind leg. The thumb is free and has a sharp claw, and is used by the bat in climbing and to hang himself up when he goes to sleep. The wings of the bat and the flying reptile have a much stronger resemblance to each other than either has to the wing of a bird. Then, just as some of the early reptiles returned to the sea, turned their legs into paddles and even acquired an almost fish-like form, so several quite distinct mammals have done the same. The whales are descended from land mammals with four feet, yet now they have the spindle-shaped body so suited for easy travel through water, their hind limbs have disappeared except for mere vestiges, and the forelimbs are turned into flippers, though they retain in most cases the same bones and joints as other mammals, and five fingers.

YELLOW-CROWNED PENGUIN'S FLIPPER ARMS

Birds having turned what were once fore-legs into wings, some of them, like the penguins, have gone through yet another transformation and made their wings into flippers for swimming. This yellow-crowned penguin of New Zealand has clawed feet and very thick "ankles."

Then there are the sea-cows (*Sirenia*), which, though unrelated to the whales—their nearest relatives alive to-day are probably the elephants—have undergone a similar change of form and of limb. The elephants themselves, though in many respects among the most specialised of mammals, have been very conservative in regard to their feet; they have retained the general form and arrangement of the bones and they have not lost a single toe, though, as the bones of the toes are comparatively short and very massive in proportion to their length, as befits the great weight they have to carry, they do not show externally except by the five hoof-like nails.

There have been among mammals three important branches or lines of evolution—the herbivorous or grass-eating animals, the flesh-eaters, and a branch which, refraining from over-specialisation and holding a nice balance of potentialities, placed their hope of survival in adaptability and a growing brain, and so led through humble lemur and monkey-like forms up to man. The life habits of the grass-eaters tended towards the pacific. They might have, and indeed some of them did, acquire such an equipment of strength and of defence as to make even the fiercest beast of prey respect them, but the tendency of most

Neville Kingston

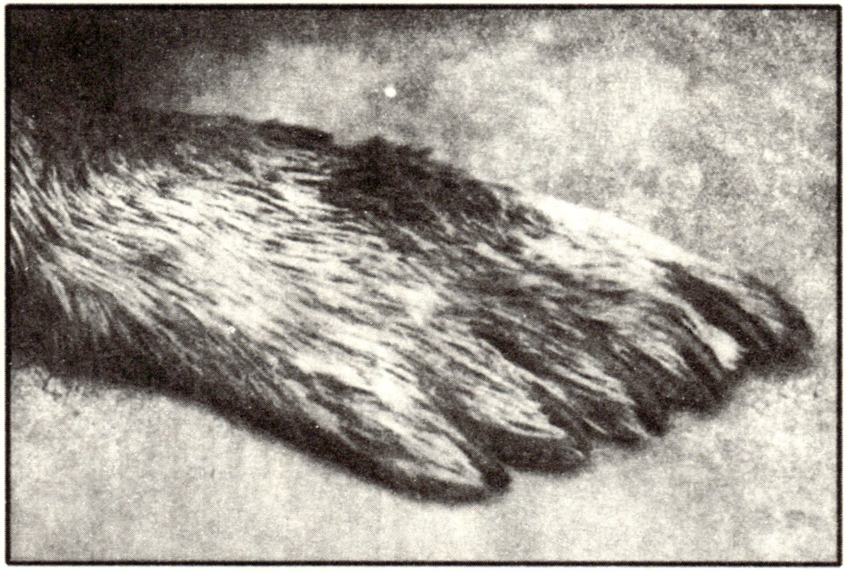

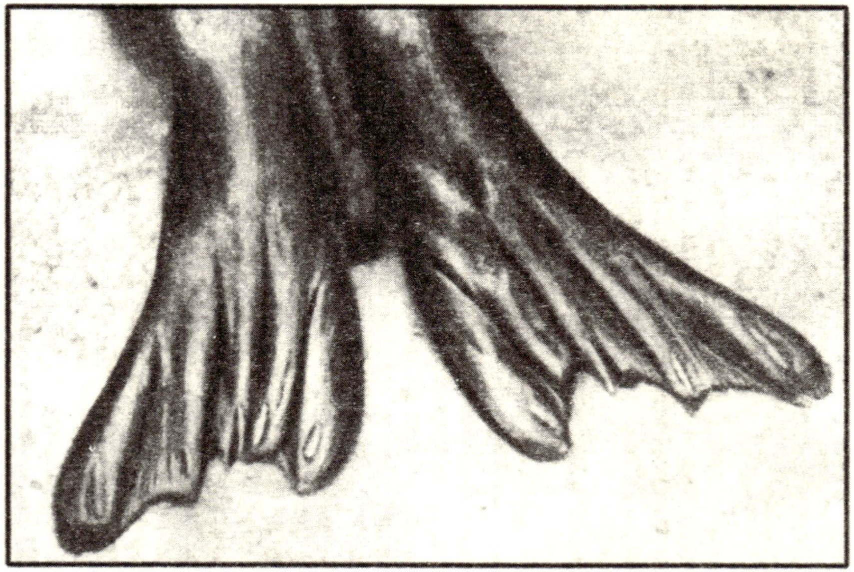

W. S. Berridge

SEAL, SEA-ELEPHANT AND WALRUS AND THE STRUCTURE OF THEIR SPECIALLY ADAPTED LIMBS

In the true seals, which are the descendants of land carnivores which have betaken themselves to the sea for a living, the change from foot to flipper has gone a long way. The fore limbs (bottom left) have the toes webbed, but have retained strong claws. But the back limbs (bottom right), have become quite useless for walking and are permanently directed backwards, being of use only in the water. The sea-elephant or elephant seal (top left) has the same characteristics as regards its limbs while the walrus (top right) has its hind limbs turned forward beneath the body. We see the animal here scrambling out of its swimming pool at a zoological garden and displaying its clawed fore-flippers.

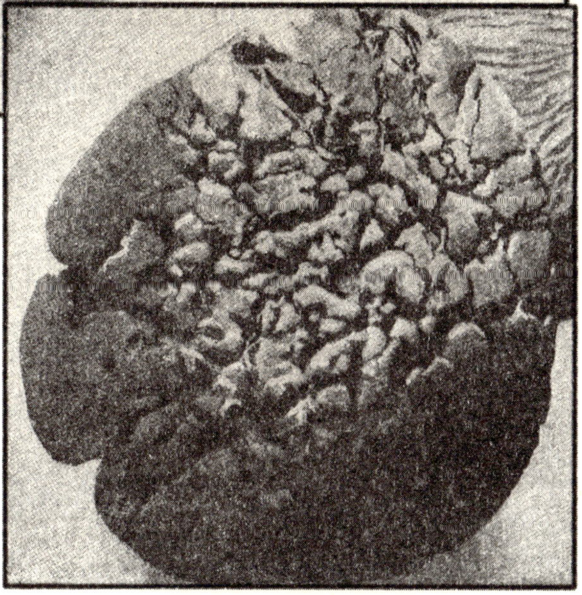

E.N.A.

ELEPHANT'S FOOT AND ITS NAILS

Below we have a remarkable photograph of an elephant's foot, taken from beneath. The toes are so short that they do not, save for the nails, protrude through the flesh. Above is an elephant in the Bronx Park Zoo, U.S.A., having its nails filed.

of them was to evade danger by fleeing from it. So, in the evolution of their limbs, everything was sacrificed to speed. They rose on their toes—every one has to do that in running—the bones of the thigh and upper arm became so overlaid with muscles and flesh that they are almost merged in the body, and so the knee and elbow joints are almost hidden, and what we call the knee of a horse is really its wrist, while the prominent joint on the hind leg (hock) is really the ankle. The long bones (called the cannon bones) beneath the horse's "knee" and hock correspond to the bones of the palm of our hand and our instep; but instead of having several bones side by side as we have, the horse has got rid of all but the centre one, which is of great thickness. The horse has also lost all his toes and fingers except the middle one on each foot, which, like the bone above it, is immensely strengthened. Moreover, the horse walks almost on the end of his toe, and to protect it the nail has grown into a kind of horny box which we call the hoof and protect further with a steel shoe.

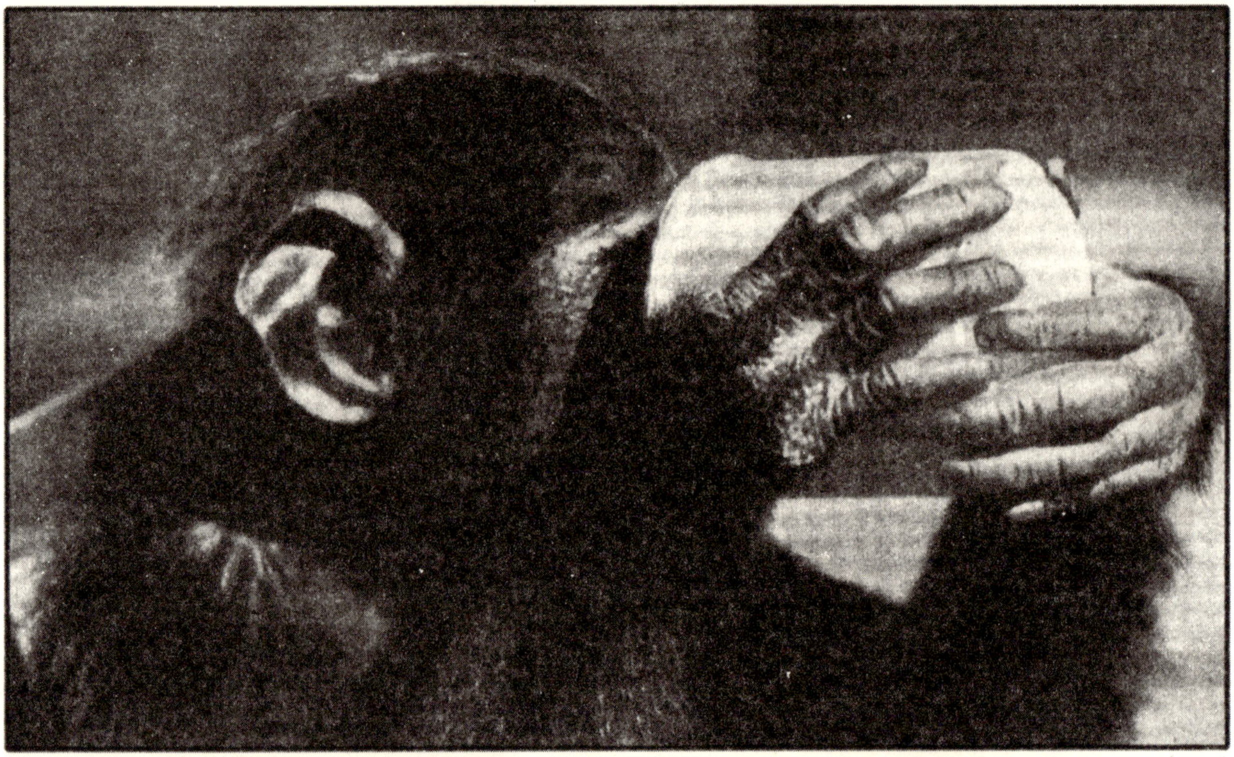

F. W. Bond

E N A.

GORILLA, BABOON AND CHIMPANZEE USING THEIR HANDS

When, in considering the scale of evolution, we have got as high as the monkeys, we find that we have got very near to the true hand. The birds started the process by turning round one of the digits of the hind limbs so that its action could be opposed by the others thus enabling a grip to be taken. Not all monkeys have a thumb, but most of the Old World monkeys have one. Here are examples of the uses of monkey hands, the gorilla, for instance (bottom left), in walking and the mother baboon (bottom right) in carrying.

Autotype Co.

WONDERFUL AND MINUTE HOOFS OF THE LITTLE KLIPSPRINGER

The name of this little antelope which lives in East Africa is klipspringer, and this signifies "rock-jumper," for it has an amazing power of jumping from crag to crag. The hoofs are so small that if all four are placed together they can stand on a five-shilling piece. The antelopes, along with goats and deer and pigs, have discarded all but two of their toes for practical use, and the hoofs are really the nails of these toes. In some cases the remnants of two other toes can be seen.

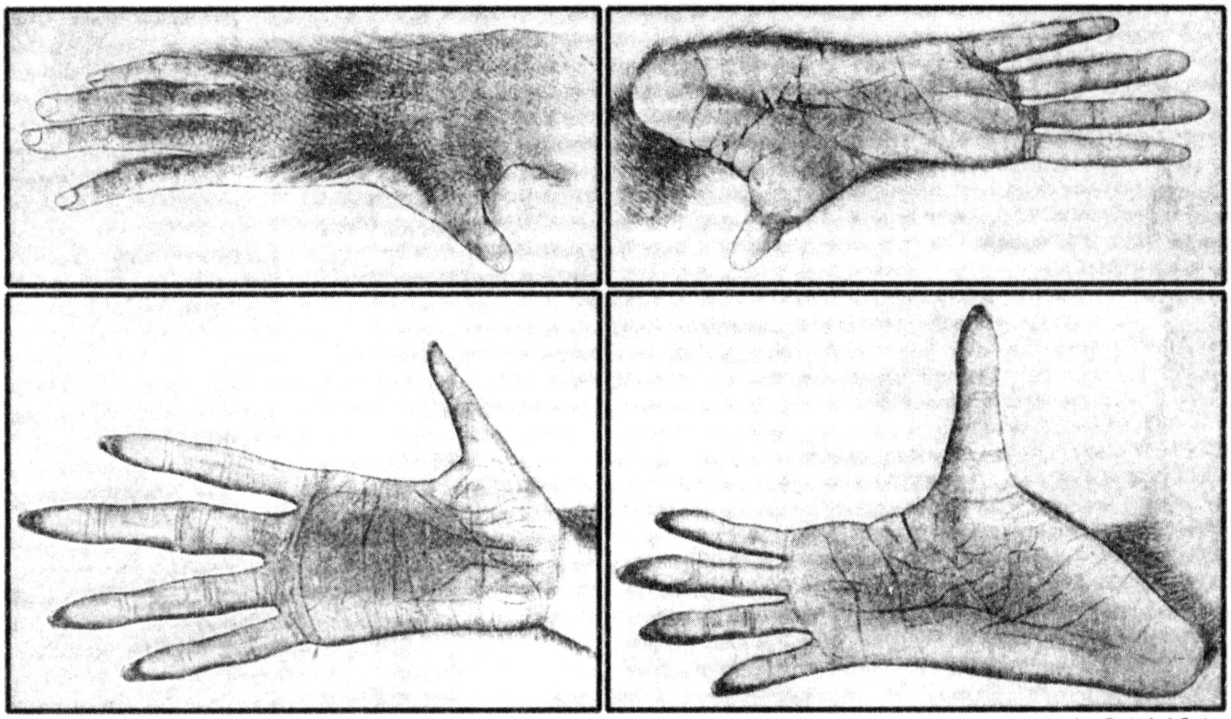

London Zoological Society

MONKEY HANDS AND FEET IN DETAIL: CHIMPANZEE AND ORANG UTAN

These hands and feet show the development attained by the chimpanzee and orang utan. The left-hand bottom photograph is of the hand of a chimpanzee and shows the large palm and fingers, and the comparatively small thumb, while the foot of the same animal (bottom right) shows the great toe developed like a thumb. The orang utan has a rather atrophied big toe as seen from the back (top left), while the other toes are long and the front view (top right) shows the palm-like surface of the foot.

A similar kind of change took place in other groups of grass-eaters—the cattle, antelopes, sheep, goats, deer, camels, and swine—only instead of throwing away all but one toe they retained two on each foot of equal length, which, encased in the box-like nail-growth, are the "cloven" hoofs of these animals. In most of them the ends of two other toes can be seen, much reduced in size and covered by little hoofs which do not reach the ground.

In the flesh-eating group of mammals we find four kinds of foot. The bears, which are carnivorous more by relationship than by habit, since they live as much on vegetable food as on flesh, are conservative people who have kept to the old custom of walking on the soles of their feet, and the bones in a bear's foot correspond very closely with those of our own. In the more truly carnivorous animals of this group there has been a certain conflict of function—on the one hand a call for speed to catch the swift prey and on the other a need for grasping power to hold it when caught.

The seals and sea-lions are the descendants of land carnivores which have returned to the sea and have turned their paws into flippers. They have kept the full number of fingers and toes and have grown a strong webbing of skin between them. In the sea-lions the limbs still possess a considerable degree of freedom of movement, and a sea-lion can travel on land at a very respectable speed with a galloping motion. In the case of the seals proper, however, the change of foot into flipper has gone further; the hind limbs project behind and are useless except for swimming, and these animals can move on land only by a laboured, awkward shuffle.

In the monkey folk the true hand is drawing very near. The first step towards it was one already taken by the birds, the turning round of the first digit or thumb until it could be opposed to the remaining fingers and so enable the hand to close on an object. The monkeys, however, did not all do this; some living species indeed have lost their thumbs altogether and use their fingers chiefly as hooks with which to hang on to the branches of trees.

Ordinary monkeys still walk on all-fours, with their feet and hands flat on the ground; nevertheless, they are capable of grasping things in true hand fashion. The chimpanzees and other apes use their hands in walking on all-fours, but they usually do so by resting the back of the fingers or knuckles, not the palm of the hand, on the ground, and they often assume a stooping erect position resembling the attitude of a decrepit old man.

Last stage of all came man himself, of humble pedigree, with a foot not much better than a bear's, and a hand not so materially different from an ape's, but with upright carriage, hands set free completely from their former function as feet, and with the enlarged brain and growing mind that enabled him to use them to the utmost and to which, alone, he owes the supremacy which he has gradually attained over all other animals.

Chapter LXXXI

Masons and Builders on Land and Sea

By W. S. Berridge

Author of "Marvels of Natural History"

In the study of animal life one cannot fail to notice how various creatures often display a tendency towards following some human vocation. Many of them, for instance, are expert trappers and hunters; some follow the unromantic but nevertheless useful occupation of a scavenger, while others, to which we propose to confine our remarks, are builders and masons which frequently exhibit such ingenuity in the construction of their homes that it is difficult to realize the work they accomplish is merely the outcome of instinct, and not of reasoning power.

Although the birds are the most familiar builders of the animal world, certain mammals, reptiles, amphibians, and invertebrates also follow the example of the feathered folk and make nests, the structures sometimes being very elaborate.

The gorilla, the chimpanzee, and the orang-utan build rough platforms of sticks amidst the branches of the trees, and upon these hard beds the females and young ones sleep, while the males take up their stations below so as to be ready to defend their families in case of need.

Some years back an adult orang utan at the London Zoo escaped from its cage after dark, and made its way into the grounds. The ape had spent the greater part of its life in captivity, and was accustomed to sleep in the quarters provided for that purpose, yet no sooner had it secured its liberty than it climbed to the top of a tree and proceeded to make a nest—an incident showing in a remarkable manner how retentive an instinct may be.

It is somewhat curious that, although the monkey family comprises such a vast and varied number of members, nest-building is only practised by those higher types already mentioned, and a few of the lower forms, as exemplified by certain species of mouse-lemurs and galagos, the arboreal homes of which are composed of twigs, leaves, and grasses, and sometimes lined with hair.

But closely allied to the lemurs is a creature, known as the aye-aye, which comes from Madagascar. It makes a large ball-shaped nest measuring about two feet across, the exterior consisting chiefly of the leaves of the traveller's tree, while the interior is bedded with small twigs, grass, and so on. The natives hold the creature in great dread, and refrain from molesting it. They believe it possesses the power to destroy those that seek to do it an injury, and should a man inadvertently catch one in a trap set for some other animal, he liberates the captive, after smearing fat over its body as a peace offering.

Of British mammals we have in the little harvest mouse an animal that has earned considerable fame for the beautiful nest it constructs. Attaining but one-sixth of an ounce in weight, and measuring only two inches in length (exclusive of its tail), this tiny creature weaves blades of grass and other material into a globular-shaped structure about the size of a cricket ball. Sometimes this is suspended from a twig, or it may repose in the head of a large thistle, but more frequently than otherwise it is bound to and supported by several stalks of growing corn. The nest only serves as a summer residence, and when the weather becomes chilly the mouse retires underground, where it remains in a torpid condition throughout the winter.

The common dormouse is another animal of our countryside that builds a nest, but instead of being content with one, it makes two, the first being utilised as a summer residence wherein the young are born and reared, and the other serving as winter quarters and a storehouse for food. These shelters, which are globular in shape and composed of leaves and moss, are built amidst hedges and bushes.

The European squirrel dormouse and the white-footed mouse are yet further examples of the diminutive nest-building rodents.

The former is an animal that used to be reared in captivity by the Romans for serving at their banquets, but the taste for such a dainty appears to have died out, and we very much doubt if nowadays even a Soho restaurant could produce such a dish should the request arise. The most likely place to find the creature would be in the cavity of an old tree, though sometimes it appropriates the deserted home of some other animal, patching it up and otherwise making it fit for its own habitation.

The white-footed mouse, on the other hand, often builds its nest of moss and strips of bark, suspended from the branch of a tree, as much as fifteen feet above the level of the ground. Possibly it finds such a situation more bracing and more conducive to sleep than one at a lower elevation.

The building habits of the musk-rat, or musquash are of interest. It makes its headquarters in a burrow containing a large sleeping chamber, the entrance to which is situated beneath water. But often it will erect a special winter domicile composed of grass and weeds plastered together with mud, the foundation resting upon the bed of a stream or swamp, and the material being piled up so that the domed roof is well above the level of the water. This "hut" or "house," as it is called, is frequently of considerable size, and as the herbage with which it is built is of an edible nature, the residence serves both as food and shelter for the inmates.

The squirrels, the marmots, and the beavers are other rodents that make nests. Some of the former dwell in underground homes, but others, such as the British squirrel, build large arboreal shelters or "dreys"

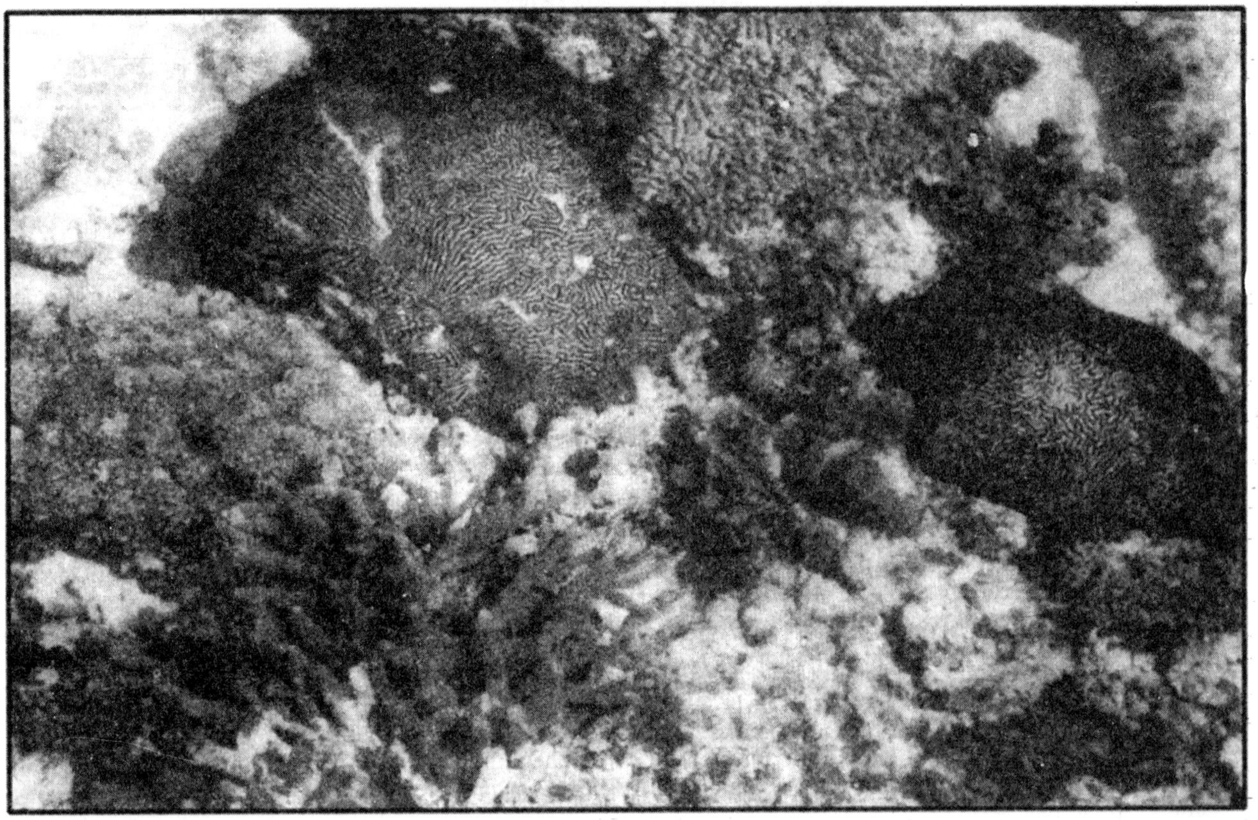

Otho Webb

W. Saville Kent

COLONIES OF MINUTE SEA CREATURES THAT BUILD WONDERFUL HOMES

In the lower photograph we see a beach at Port Denison, Australia, which is covered with lumps of coral which have been broken from their parent reefs by storms and washed up by the waves. The damage done to the reefs whence these fragments come will soon be repaired by the wonderful little creatures which, out of carbonate of lime extracted from the sea water, fashion dwellings for themselves which exhibit all the hues of the rainbow. The top photograph shows the variety of coral that may be met with in a few square yards

Martin Duncan

W. S. Berridge

A CORAL "CLOSE-UP" AND SOME PLANT-LIKE FORMS

In the lower photograph there is a very close view of a piece of coral showing the surface built up, seemingly, in a series of layers, and pitted with a number of minute holes. The upper illustration shows a group of reef-forming corals whose forms are distinctly plant-like. The coral polyps build all this from the lime in the sea water.

The ground thus heaped up not only serves to prevent the rain from flooding the home, but it also constitutes an elevated perch from which the animals can scan the surrounding country on the look-out for lurking foes.

It is, however, the work of the beavers that calls for special notice, for the constructive ability of these creatures is more highly developed than that of any other mammal, combining as it does the craft of mason, builder, and carpenter.

Although sometimes making a simple burrow in the bank of a stream for a habitation, more frequently than otherwise beavers erect in the midst of the water a large structure or "lodge" composed of twigs and branches, cut into suitable lengths with the aid of their chisel-like teeth, and held together with a plastering of mud. One or more compartments are situated within the lodge, and a bedding of wood-chips, leaves, and so on, is added to make the home

composed of leaves, fibrous roots, moss or dead bracken. These have a domed roof, and the entrance is often made more attractive and weatherproof by the addition of a projecting eave.

The marmots all excavate underground burrows as a habitation, and vast areas of ground are often undermined to such an extent as to prove a source of danger to riders on horseback. Tunnelling with the aid of their forefeet, the animals loosen the soil and kick it backward so that it forms a cone-shaped pile around the entrance of the subterranean quarters.

comfortable. Access to the interior is obtained only through tunnels leading from below water.

But in addition to constructing these lodges, beavers will often erect a barrier or "dam" across a stream for the purpose of holding up the water and keeping it at an even level. Gnawing through and felling small trees, and collecting branches, sticks, roots and grasses, they carry or float this material to the desired position, weighing it down with stones, and cementing it together with mud so as to render it impervious. As a dam may measure

W. Saville Kent

LOWLY BUILDERS OF MIGHTY STRUCTURES IN THE OCEAN CORAL OF AUSTRALIAN WATERS

To the general public coral is an ornament. To the mariner it is often a deadly menace and many a good ship has lain with bleaching ribs on the deadly reefs of masonry which the coral polyps build up. To the naturalist these reefs wear quite a different aspect. They are to him an ever-expanding monument to the activities of a wonderful society of tiny creatures which build dwellings for themselves and, dying, leave their mansions, coloured like a flower bed in a garden, for the wonder and admiration of the men with eyes to appreciate them. It must be remembered that though coral is tremendously hard the coral polyps are soft and fragile creatures. Here we are on the wonderful Fringing Reef of Port Denison, Australia.

as much as two hundred feet in length, and be of considerable thickness at the base, one can well imagine the amount of labour expended in its erection.

It has often been stated that beavers employ their tails as a mason's trowel for the purpose of plastering the mud upon the face of their dams, but this is but one of the unnatural natural history stories that are continually being repeated, and has no foundation in fact, for the creatures merely use their fore-paws for that purpose.

No account of mammalian nest-builders would be complete without mentioning the mole. Its underground home or "fortress" is a very elaborate affair, consisting of a central chamber, bedded with grass and leaves, from which numerous galleries radiate in various directions. Some of these do not extend far, but that known as the "bolt" or "main run," which is of greater circumference than the others, is directed downwards from the central chamber, and then turns upwards and connects with a tunnel having an outlet to the field above.

The mole is well adapted for the life it leads, for not only are its forefeet armed with powerful digging claws, but its velvet-like fur is embedded vertically in the skin so that it does not become ruffled or clogged with dirt, no matter whether it be brushed backwards or forwards. The creature is a model of industry, and during its foraging expeditions has been known to tunnel through the earth for a distance of 450 times its own length during the course of a night.

Kangaroos are probably the last creatures that one would expect to make nests, but some of the smaller kinds, known as rat-kangaroos, are in the habit of fashioning such shelters. Scraping a depression in the earth beneath a log, or in the midst of thick undergrowth, the animal then collects bunches of grasses, reeds, etc., carrying the material back to the site of the nest, not with its fore-paws, as one might well expect, but with the aid of its prehensile tail. When all is complete, the kangaroo takes a header beneath the pile of litter, covering itself up so as to be out of sight of its enemies.

Another very interesting nest-builder is the platypus, or duckbill, an aquatic mammal found in Southern and Eastern Australia, as well as in Tasmania, but now on the verge of extinction. The animal is remarkable in the fact that the female lays eggs from which her young are hatched, the parent sitting upon the ova during the period of incubation, in the manner of a bird. Chiefly nocturnal in habits, and of a very shy disposition, the platypus makes its headquarters in a burrow excavated in the bank of a stream. Tunnels leading to this home sometimes extend for a length of fifty feet, and entrances are usually situated both below water and above ground.

Constructive ability is revealed to a very high degree in many birds, and when one compares their mental and physical development with that of human beings there is just cause to wonder at the varied accomplishments achieved by the feathered folk with but a minimum of material. Pages could easily be devoted to dealing with the building habits of birds, but we must confine our remarks to a few of the more striking examples.

Let us then take as our first subject the curious hammer-head, or tufted umbre, an African bird about the size of a duck. It is imbued with somewhat original ideas in regard to what should constitute a desirable residence, for instead of being content with a one-roomed establishment, it prefers a bijou flat containing several compartments. Such being the case, the nest must necessarily be of considerable size. Not infrequently it measures more than six feet in diameter, and its strength is so great as to be capable of withstanding the weight of a man sitting on the top. Dome shaped, and composed chiefly of twigs, this "ideal home" consists of three rooms that may be likened to the entrance hall, the day-nursery, and the night-nursery of a human habitation. The latter, which is on the upper floor, is where the young ones are hatched and reared during the early days of their existence; the second or middle room serves as supplementary accommodation when the family grows up and requires more elbow-room; while by standing in the "hall" and looking out of the "front door" the inmates are able to see what is going on outside without exposing themselves unnecessarily to observation.

There is a number of birds which construct mud nests, the swallow and the house martin being familiar examples of such avian masons found in this country, while the oven bird and the grey struthidea may be mentioned among those from foreign climes.

The mud nests of both the swallow and house martin are lined with dried grass and feathers. As a rule they are attached to a wall and placed under the eaves of a house, but it is interesting to note that the former bird has been known to build in a tree, while Gilbert White mentions an individual that for two successive seasons built on the handles of a pair of garden-shears resting in an outhouse, and another that made use of the dried body of an owl, hanging from the roof of a barn, as a support for its nest.

But although the swallow and house martin exhibit such ingenuity in fashioning their homes, the South American oven bird is an even more expert builder. The bird is no larger than a thrush, yet it erects a nest weighing as much as nine pounds, and measuring twelve inches across the widest part. Made of mud, combined with fibrous roots and horse-hair, the walls of the structure are half an inch in thickness, and become baked to the hardness of a brick by the heat from the sun. No attempt is made at concealment, the nest usually being situated in a very conspicuous position, such as upon the top of a post, on a large slab of stone, or on the leafless branch of a tree. A small hole serves as an entrance to the home, and inside there is a chamber partitioned off by a wall of hardened clay extending nearly across from one side to the other.

G. Hearn

SWALLOW'S NEST IN A BOX

Under the eaves is the usual and indeed the traditional nesting site of the swallow, but this bird sometimes becomes freakish with regard to its accepted habits. Here is a swallow which built its mud nest in the corner of a packing case.

Also composed of mud, and lined with grass, the nest of the grey struthidea, a bird found in Southern and Eastern Australia, is a curious, bowl-shaped erection, open at the top. But the most remarkable fact concerning this tenement is that but a very small portion of the lower and flattened surface is in contact with the bough of the tree upon which it is built, the greater part being entirely unsupported.

Yet another example of a bird that utilises mud in the making of its nest is the hornbill. Selecting a cavity in a tree, the hen lines the bottom with chips of wood, earth and feathers. She then takes up her residence in her "little wooden hut," and settles down to parental duties. The male plasters up the entrance with clay, leaving only a small, slit-like orifice through which his wife may thrust her beak and receive her daily ration of food. Although the prisoner must have a rather trying time whilst thus shut up in such confined quarters, she at least has the satisfaction of knowing that her young ones will be safe from the unwelcome attentions of those creatures that would be only too willing to sup upon them should the opportunity arise.

But possibly the strangest of all birds' nests are those of the edible swifts, for they are composed almost entirely of the secretion from the bird's salivary glands, which becomes hardened upon exposure to the air, and looks like isinglass. Shallow, saucer-like structures, and attached to the face of the caves whither the birds resort in considerable numbers, the nests form the chief ingredients of the bird's-nest soup so beloved by Chinese epicures. Owing to the great demand for these dainties, many are destroyed and collected as soon as they are completed, but the industrious workers immediately commence to build new ones, although feathers and other suitable materials are then intermingled with the mucus so as to make up for lost time. In former days it was thought that the substance constituting the nests was yielded by seaweed or some other vegetable matter collected by the birds, but analysis has revealed without a doubt that such is not the case.

VERY notable exponents in the art of nest building are the weaver birds, the tailor birds, and the Australian flower-peckers. The baya, or common weaver bird fashions a retort-shaped structure composed of strips of grass and palm leaves, coconut fibre, and so on, woven together in a most wonderful manner. The lower part of the nest is extended into a long, tubular neck, the extremity of which is left open so as to form an entrance. Attached to the top is a woven cord, the free end being fixed to the bough of a tree, or some other support. The upper part of the entrance-tube is partially screened off internally by a crosswise partition, this being a safety device to obviate the risk of any unfledged birds tumbling down the spout. The cock and hen birds both share in the making of their home, and frequently one will work inside the half-finished nest while the other remains outside, the two passing the ends of the grasses in and out to one another, and pulling them tight.

The sociable weaver birds construct a different type of nest from that of the baya bird. As implied by their name, they are of a very sociable disposition and like to dwell in close proximity to one another. Their communal home is a large, solid, umbrella-shaped structure erected amidst the branches of a tree, the under part being flattened and pierced by numerous holes, each of which denotes the entrance to a nest.

That truth is stranger than fiction is well exemplified in the habits of the tailor bird. Employing its beak as a needle, it actually sews together with vegetable fibre, or the silk from the cocoons of caterpillars, the edges of several adjacent leaves, leaving a small opening at the lower end as an entrance. Within this envelope the proper nest is built with down and fibre.

Equally wonderful are the nests of the flower-peckers. Composed of spiders' webs and the down collected from seed pods, these fairy-like habitations are so soft, and yet so strong, that it is possible to fold them up without their being broken.

QUAINT "MUD-PIE" NESTS OF A COLONY OF FLAMINGOES

The nests of flamingoes vary between a kind of soup-plate shape in those built on the mud flats and marshes, and a cone when constructed actually in the water. Here are some photographs of such colonies of nests as these birds fashion. Below we see some of the quaint parents, and, in the left foreground, two are standing on their nests and the single egg can be seen in the hollow of the nest. Above is a deserted colony in the Bahamas. The nests are made high to avoid being flooded during the rainy season, when they are built.

W. S. Berridge

WELL-BUILT HOMES CONSTRUCTED BY CLEVER MASONS AMONG THE BIRDS

One of the most notable of bird masons is the oven bird of South America. Though only about the same size as a thrush, it builds itself a home (bottom left) weighing as much as nine pounds and measuring a foot across. This weighty structure is not concealed in any way, but is erected upon the branches of a dead tree or on a rock. The materials are mud, horsehair and roots, the walls being half an inch thick and baked hard by the sun. The Grey Struthidea of southern and eastern Australia makes a bowl-shaped nest out of mud (bottom right). The other photographs are of a martin and its nest under the eaves (top left), and the open, cup-like structure of the swift which builds the nests from which the Chinese make their famous bird's-nest soup.

Quite a different kind of avian constructional work is exhibited by the bower birds, which erect bowers, or playgrounds during the mating season, and before the commencement of nest building.

The gardener bower bird builds a conical-shaped, hut-like edifice of twigs, about two feet in height, and covered with moss. A portion is left open, and in front of this doorway there is a cleared space, bedded with more moss, and decorated with flowers, brightly coloured insects and beetles. These are replaced when they lose their brilliance. The satin bower bird makes its bower in the form of an avenue, embellishing it with shells, flowers, and other brightly coloured objects; while Newton's bower bird erects around the base of two closely adjacent and slender tree-stems a heap of twigs that may in time get piled up to a height of as much as nine feet. It has been noticed that one pillar is always higher than the other, and that only the taller of the two is decorated.

Giving our attention to the fish that make nests, we have in the sticklebacks well-known examples found in British waters. The three-spined stickleback, or "tiddler" collects floating pieces of fibrous roots, straws and grasses, and, transporting them to the bottom of a shallow stream, makes a muff-shaped nest, open at one end, by binding together the material with mucus threads secreted by his kidneys. The female deposits her eggs inside this shelter and then departs, and when the young ones hatch out the male pulls to pieces the roof of the structure, thereby converting it into a kind of cradle.

Very similar is the nest of the ten-spined stickleback, although it does not rest upon the ground, but is woven around and supported by the stems of water plants; while that of the fifteen-spined stickleback, or "marine adder," is a pear-shaped structure composed of fronds of seaweed which the male binds together with the aid of a pliable thread, exuded from his mouth, that quickly hardens in the water.

Another curious form of nest is that made by the tentacle fish. It builds in the midst of the gulf-weed of the Sargasso Sea, pieces of seaweed being joined to one another by a pasty secretion yielded by the fish, and the large, grape-like bunches of eggs laid by the female being suspended therefrom.

The fresh-water fish from the United States, known as the bowfin, makes a nest amidst thick reeds and aquatic vegetation. The male bites through and removes the stems of reeds and plants, thereby clearing a circular space with a passage-way leading to the open water beyond. On the floor of this prepared area he scoops out a saucer-shaped depression to a depth of about eight inches, and with a diameter of several feet, remaining in the privacy of this sanctum for the arrival of the females to lay their eggs.

But the most remarkable of fishes nests are probably those of the paradise fish and the fighting fish. The males blow bubbles from their mouths and these rise to the surface of the water and form a mass of froth. Each nest is about one quarter of an inch in depth, and covers an area of about six inches across the widest part. When these floating abodes are finished, the females take up their position beneath and liberate their spawn, the eggs being so light that they rise upwards and settle amidst the froth. In this situation the young ones hatch out, being guarded by the males.

The only other creatures that make bubble nests are certain kinds of frogs. Some of the piping frogs excavate a small hole in the ground, and then fill the cavity with a frothy mixture that keeps the spawn moist; while the nest of White's tree frog is composed of bubbles, and floats upon the surface of the water. Very similar are the habits of a flying frog found in Japan, for the jelly-like substance with which the eggs are enclosed is beaten up into froth by the creature's hind legs.

Crocodiles and alligators must be added to our list of nest builders. The female of the Mississippi alligator makes a large mound of leaves, twigs, and grasses, the heap measuring as much as ten feet in diameter, and several feet in height. She deposits her eggs amidst this material, covering them up with moist vegetation, the decaying of which generates a considerable amount of heat and hatches the young.

But the nest of the crocodile is a much simpler affair than that of the alligator, and consists merely of a depression scooped out in the sandy soil. From forty to sixty eggs are laid therein, after which they are covered with sand and usually left to be hatched by the heat from the rays of the sun. Sometimes, however, the female will assist in the process of incubation by resting upon the place beneath which her eggs repose.

Of the lower forms of life that follow the occupation of a mason or builder, brief mention must be made of the corals, bees, wasps, ants, and termites.

It is not always realized that the hard, stony substance of commerce called coral is nothing more than a deposit of carbonate of lime extracted by the living creatures from the sea wherein they dwell, the organisms themselves being soft-bodied and jelly-like. Some corals are separate and of simple structure, but others, as exemplified by the reef-builders, are what is termed compound—that is to say, they increase from the parental stock by a continual process of budding, each new individual being connected by tissue to another, so that the entire colony, no matter how large it be, may be likened unto a single living mass which possesses innumerable mouths and digestive organs.

In the Pacific and Indian Oceans, far distant from land, large masses of coral arise from the depths to the level of the water, and often extend for a distance of many miles. These reefs consist of living coral at their upper and outer surfaces, the inner and lower parts of the structures being dead. Corals are frequently most beautifully coloured, those that constitute the Great Barrier Reef of Australia, for instance, varying in tint from red, purple and violet to emerald and orange.

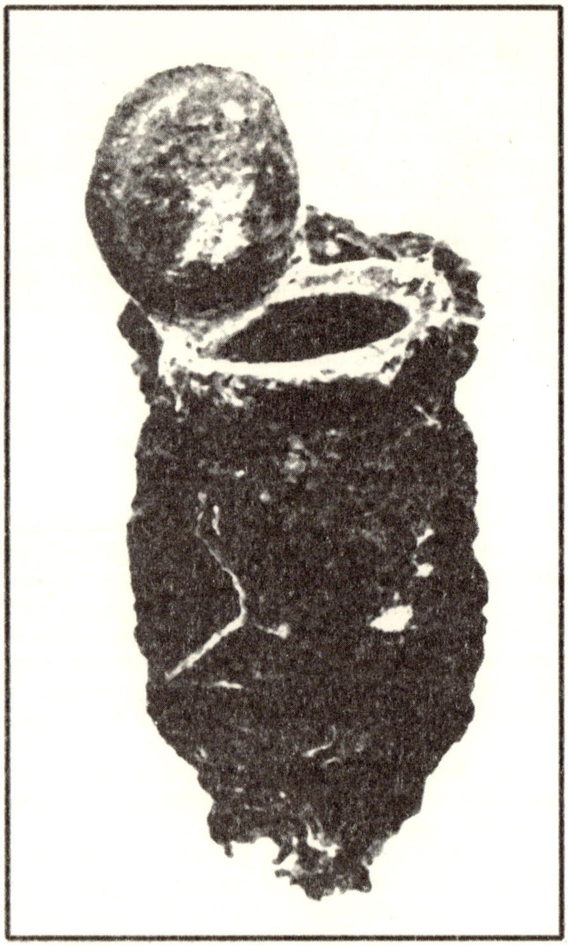

Martin Duncan

TRAP-DOOR SPIDER'S DWELLING

Certain spiders called, from their building habits, trap-door spiders, make a burrow in the earth and shore up the walls with layers of silk. The entrance is provided with a trap door or lid, built of earth and provided with a hinge of silk.

The majority of ants make nests. One kind found in South America collect and carry in their jaws fragments of earth which they deposit on the branch of a tree and fashion into a home perforated by tunnels leading to dwelling chambers. The latter are lined with a papier-mâché-like substance consisting of masticated wood-dust mixed with a tenacious fluid that flows from the ants' salivary glands. The exterior of the nest is sown by the tiny workers with the seeds of various plants. These germinate and throw out leaves, the foliage then acting as a screen.

Another species of ant, from Central America, makes a nest within the large thorns protecting the branches of acacia trees; while the red or spinning ants join together a number of leaves, the edges of which are made to adhere to one another with a silky thread which is produced by the grubs just prior to their becoming transformed into the chrysalis stage.

The wood ants, the largest species found in this country, fashion a nest that takes the form of a heap of small fragments of sticks and vegetable substances, honeycombed with galleries and chambers; while the black garden ant, with which we are so familiar, often adds to the extent of its underground home by gnawing away the interior of dead branches or rotting woodwork, the material removed being chewed up into a "paper" employed in the erecting of walls and supports.

With the so-called white ants which, although very ant-like in appearance, are not ants, but should more properly be spoken of as termites, we come to some atoms of life that erect enormous nests of earth and wood-pulp, mixed with saliva, that sometimes project above the level of the ground to a height of twenty or thirty feet. The walls are extremely hard, and the entrances to the nests are always situated underground.

The bees and wasps exhibit remarkable skill as builders, and the six-sided cells of wax that many fashion for the storage of their food and for residential purposes are familiar to all.

The mason wasps and the potter wasps make their homes of earth. Two species of the former found in Britain excavate tunnels in sandy banks, the material removed being erected around the mouths of the abodes into projecting and tubular spouts that help to prevent the ingress of another kind of wasp that seeks to lay its eggs in the nests of the rightful owners.

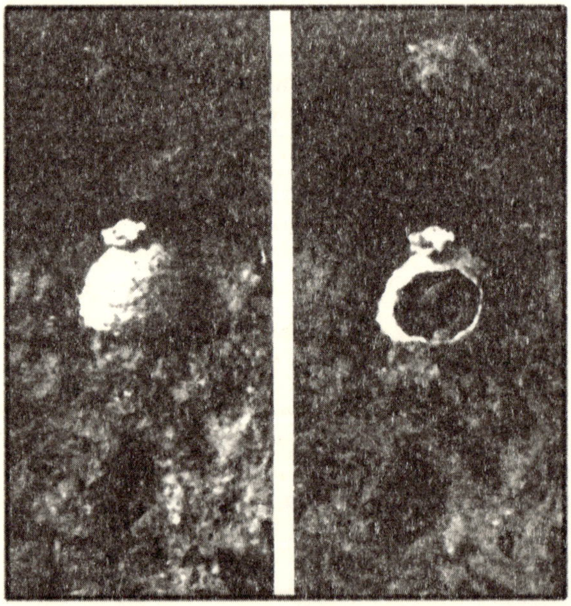

Hugh Main

NEST OF A MUD-WASP AND ITS CONTENTS

On the left is a view of a mud wasp's nest shaped rather like a bag tied at the top. On the right we see the nest opened to show the developing wasp grub inside, and a caterpillar provided by the parent for the growing grub's food.